A Seriously Humorous Voyage Through the Universe

Learn about Black Holes, Time Travel, Quarks, Spooky Action, and a Heck of a Lot More

Oscar Joe Washington

Highest Hill Publishing

© Copyright - Highest Hill Publishing - 2023 - All rights reserved.

The content contained within this book may not be reproduced, duplicated, or transmitted without direct written permission from the author or the publisher.

Under no circumstances will any blame or legal responsibility be held against the publisher, or author, for any damages, reparation, or monetary loss due to the information contained within this book. Either directly or indirectly. You are responsible for your own choices, actions, and results.

Legal Notice:

This book is copyright protected. This book is only for personal use. You cannot amend, distribute, sell, use, quote, or paraphrase any part, or the content within this book, without the consent of the author or publisher.

Disclaimer Notice:

Please note the information contained within this document is for educational and entertainment purposes only. All effort has been executed to present accurate, up-to-date, and reliable, complete information. No warranties of any kind are declared or implied. Readers acknowledge that the author is not engaging in the rendering of legal, financial, medical, or professional advice. The content within this book has been derived from various sources. Please consult a licensed professional before attempting any techniques outlined in this book.

By reading this document, the reader agrees that under no circumstances is the author responsible for any losses, direct or indirect, which are incurred as a result of the use of the information contained within this document, including, but not limited to, — errors, omissions, or inaccuracies.

Contents

Epigraph	V
1. Shedding Light on the Universe – Understanding the Nature of Light	7
2. Spooky Action at a Distance	10
3. Introducing Uncle Albert	14
4. In The Loop – The Quantum Realm	20
5. Very Special Relativity	28
6. The Terrifying Black Void – the Black Hole	36
7. Surfing the Gravity Wave	39
8. Chasing Unicorns: The Quest for a Unified Theory	41
9. Breaking the Space-Time Continuum: Understanding Time Travel	43
10. The Multiverse Theory: Are We One of Many or Many of One?	47
11. Starry Nights: Exploring the Wonders of the Night Sky	55

12.	From Bright Giants to Dim Dwarfs: Understanding the Star Magnitude Scale	60
13.	Connecting the Dots: Exploring Constellations	62
14.	The Expanding Universe	65
15.	A Journey into the Heart of Stars	68
16.	Unraveling the Mystery of Dark Matter and Dark Energy	81
17.	Quantum Theory: When Particles Don't Play by the Rules	84
18.	Discovering the Diversity and Wonder of Solar Systems in the Universe	87
19.	Interstellar Pioneers: Tracing the Epic Journey of the Two Voyager Spacecraft through the Uncharted Territories of the Outer Solar System	96
20.	Space Objects	131
21.	The Exciting Quest for Exoplanets	147
22.	Is Anybody Out There? The Search for Extra-Terrestrial Life	151

Conclusion	158
About the Author: OSCAR JOE WASHINGTON	163
Acknowledgements	165
Chapter	167

Life, forever dying to be born afresh, forever young and eager, will presently stand upon this Earth as upon a footstool, and stretch out its realm amidst the stars.

~ H. G. Wells, Science Fiction Author

INTRODUCTION

Listen up, folks, if you want to make sure you don't forget all the super duper important info I'm about to drop on ya, you better sharpen those pencils and get ready to take some notes! I've got a little surprise for ya. Think of it as a fun little quiz to make sure you were paying attention.

You can download the free pdf document of the **Quiz** using this link...

bit.ly/3MJMmzz

Ready? Let's do this!

All That Is, Was, and Will Be

Have you ever heard of the cosmic circus? Well, it's a real thing, and it's the universe! Imagine a clown car with so many acts and performers; your head will spin faster than a topsy-turvy rollercoaster ride. But the universe is not just your average big top show. It's a vast and mysterious place, full of wonder and awe-inspiring beauty, and strange and unpredictable enough to surprise even the most experienced astronomers.

Picture yourself standing on a platform in the middle of an infinite ocean. You look around, and it's just water, water everywhere! But wait, hold on to your cotton candy because suddenly, a pod of planets, stars, and black holes emerges from the depths, leaping and flipping through the air like a group of intergalactic dolphins.

Let's talk about the basics, shall we? The universe is a big, whopping everything! It's got all matter, energy, and space, an estimated 93 billion light-years in diameter, and over 100 billion galaxies, each with billions of stars. To put it in perspective, even light, which travels at an astonishing 186,000 miles a second, takes billions of years to cross from one end to the other. That's an awful lot of space!

The human brain is limited by its ability to conceptualize such large numbers and distances.

We know a thing or two about our own solar system and some properties of stars and galaxies. But the whole enchilada? It's a different story. Trying to grasp the full scope and scale of the universe is like attempting to count to a googolplex - our brains can only handle so much! By the way, a googolplex is a one followed by one hundred zeros.

Luckily, the brainiacs have come to the rescue with their fancy tools and models to help us visualize this humongous universe. They've got simulations, models, and scaling techniques up the wazoo. But even with all that, it's still a tough nut to crack. It's like finding Waldo in a packed stadium - good luck with that!

But the universe is not just a big, boring vacuum. It's got structure and organization from the tiniest subatomic particles to the largest superclusters of galaxies, all following the rules and laws that govern their behavior. And speaking of rules, there's the force of gravity, responsible for holding everything together from planets to entire galaxies. It's why we're not floating into space right now, thank goodness!

But let's not forget the main attraction - black holes! These guys are the most bizarre objects in the universe, formed from the collapse of massive stars under the force of gravity, creating a point of infinite density and zero volume known as a singularity. Anything that gets too close to a black hole, even light, can get trapped in its intense gravitational pull, never to be seen again. Yikes!

But black holes aren't all bad. They're also essential to the universe's evolution, like when galaxies collide, and the massive gravitational forces can cause black holes to merge, creating even bigger black holes. These cosmic collisions also result in the formation of new stars and the redistribution of matter throughout the universe.

Now, let's talk about the fun stuff, like the beautiful nebulae, vast clouds of gas and dust where new stars are born. They're like the colorful confetti of the universe, coming in all shapes and sizes, from the famous Horsehead Nebula to the stunning Crab Nebula.

And then there are the stars themselves, the bright beacons of light that dot the night sky. They come in a variety of sizes and colors, from the massive blue giants to the cool red dwarfs. Our very own sun is a yellow dwarf and is responsible for providing the energy that sustains life on Earth. So, you know, no big deal.

But hold on to your popcorn! If you think the universe isn't mind-bending enough, consider this: it's expanding. Every second, the distances between galaxies are getting larger and larger, increasing at a rate of tens of thousands of kilometers per second. It's like trying to keep up with a gazelle on your morning run!

And then there's the mysterious dark matter and dark energy, which make up most of the universe's mass and energy. We can't see or detect them directly, but we know they exist because of their gravitational effects on visible matter. It's believed that without dark matter, galaxies would never have formed in the first place.

But even though the universe is more complicated than my ex's dating life, scientists have managed to uncover some mind-blowing secrets. We now know that the universe is approximately 13.8 billion years old, and it all started with a massive explosion called the Big Bang. And there are like, trillions of galaxies out there - seriously, who has time to count them all?

What's even more fascinating is that the universe isn't just a random mess of stars and stuff. Nope, there are actual patterns and laws that keep everything in line. It's like the universe is a big old rule follower, following the laws of physics and sticking to constants like the speed of light. And while we still have a ton of unanswered questions, scientists are working hard to uncover all the universe's secrets.

So, why should you care about the universe and its crazy complexities? Well, besides making us feel small and insignificant, understanding the universe can give us a sense of our place in the grand scheme of things. Plus, the universe is just straight-up fascinating. From the beauty of the night sky to the trippy theories of quantum mechanics, there's always something new and cool to discover.

So, let's grab our telescopes and strap on our spacesuits because the adventure of a lifetime is waiting for us – this will

INTRODUCTION

be a wild ride through space and time. We'll explore everything from itsy-bitsy subatomic particles to massive clusters of galaxies. We'll dive into the mysteries of black holes, dark matter, and the Big Bang. And we'll do it all with a sense of humor and wonder that'll make you say, "Whoa, dude, the universe is cray-cray!" As the great astronomer Carl Sagan once said, "Somewhere, something incredible is waiting to be known." Let's go find it together, folks!

SECTION ONE

Time and Space, Relativity, Gravity

Chapter One

Shedding Light on the Universe – Understanding the Nature of Light

According to science, light is just a fancy way of saying that electrically charged particles like electrons and protons are getting cozy with each other. These particles create something called electromagnetic fields, which basically means they generate and transmit different types of radiation like radio waves, microwaves, and even gamma rays. So basically, light is just another electromagnetic wave, but it's special because we can actually see it with our own eyes.

So you wanna know what a light wave is? Well, my friend, it's like a rockstar of the electromagnetic world. It just loves to oscillate in a wavy motion as it zips through the vacuum of space at the speed of light. Think of it like a high-energy dance party, with electric fields jumping up and down and magnetic fields getting their groove on in a perpendicular direction.

But don't be fooled by its flashy moves because light waves are serious business. Their wavelengths determine their color, so it's like a rainbow on steroids. You've got your reds, your blues, your greens, and everything in between. And let's not forget about the frequency of the wave determines how much energy they've got. High-frequency light is like a double-shot espresso, while low-frequency light is like a lazy Sunday morning.

But don't underestimate the power of light, my friend. It's not just for making things look pretty. Light is essential for life on Earth, providing the energy needed for photosynthesis (the process where plants turn sunlight into food). It also helps us see, which is a pretty handy skill to have when you're trying to avoid getting eaten by a predator.

And let's praise the scientists out there using telescopes and other fancy gadgets to study the universe. Without light, we wouldn't know a thing about stars, planets, galaxies, or black holes. So let's give it up for light, the unsung hero of the universe. It's truly electric!

Light waves are produced by all kinds of sources, from the sun to your lightbulbs, and they can travel through anything from space to your favorite pair of sunglasses. Depending on the material they're dealing with, they can either get absorbed, transmitted or reflected when they hit something. It's like a game of ping-pong, but with light.

This little bugger called light moves faster than Usain Bolt on steroids - clocking in at a whopping 186,282 miles per second. This speed is a fundamental constant of the universe and is crucial to physics nerds everywhere. And before you ask, no, it didn't just get up one day and decide to run like the wind.

The reason light moves so darn fast has to do with the fabric of space and time itself. You see, space and time are like a dynamic duo that works together to keep the universe ticking. And it just so happens that the speed of light is the magic number that governs the relationship between these two cosmic superheroes.

It's kinda like a universal speed limit, but instead of getting pulled over by the space cops for going too fast, you just... can't. No matter how fast you're going or where you are in the universe, the speed of light is always gonna be the same. It's like the one constant in an ever-changing, chaotic universe.

But don't just take my word for it - this stuff has been proven by countless experiments over the years. And it's not just some wonky science-fiction nonsense, either. This constant speed of light is a crucial component of Albert Einstein's theory of relativity, which is like the granddaddy of all physics theories.

The speed of light is like the superstar athlete of the universe, breaking records and defying expectations. It's the glue that holds the fabric of space and time together, and it's super important to our understanding of the cosmos.

Chapter Two

Spooky Action at a Distance

Have you heard about this wacky quantum thing called entanglement? It's like a BFF bond between particles, where they can gossip with each other over long distances, just by being watched. It's like they're speaking in code, and by spying on them, we get to be in on the juicy details. Even Einstein thought it was freaky-deaky, calling it "spooky action at a distance." I mean, who needs a telephone when you have entangled particles? Call me, maybe?

Let's take a ride into the strange world of quantum mechanics! You see, back in the day, in 1935, ol' Albert Einstein and his science buddies discovered something truly mind-bending. They found that if you linked up two quantum particles, they would always have each other's backs - no matter how far apart they were! The particles essentially always KNEW about the other's properties. That's right, measuring one particle would instantly tell you the state of its twin, as if they had a secret telepathic connection.

And, Yeah, that's right, Einstein called this mysterious phenomenon "spooky action at a distance!" Even the great Einstein was left bewildered by the zany ways of quantum mechanics. But, as it turns out, this "spooky action" is real and is known today as entanglement.

Schrodinger's Poor Little Cat

So, get ready for the next astounding concept. We're diving into the fantastical realm of Schrödinger's cat! Erwin Schrödinger, an Austrian physicist, conjured up this wild idea. He became known as the "father of quantum mechanics". The scientist paused, a mischievous grin on his face, "So, ready to flip some cosmic coins and test the limits of entanglement"?

Imagine a poor kitty trapped in a box with a ticking time bomb - in the form of a radioactive substance - and a vial of poison. According to the strange rules of quantum mechanics, the radioactive substance could go boom or bust at any moment, killing the cat. But hold on; it gets even wilder! Until that box is opened, this cat is alive AND dead, existing in a state of quantum superposition. Yup, that's right, Schrödinger's cat is the king of paradoxes! It'll leave you bewildered, but this is the bizarre beauty of quantum mechanics, folks.

Schrödinger's cat

Imagine two coins, one in your hand and one in mine," said the scientist with a twinkle in his eye. "If we flip them both at the same time, the odds of getting heads or tails are about 50/50, just like any normal coin flip. BUT, what if I told you these coins have a cosmic connection, and the outcome of your flip determines mine, making my flip anything but random!"

"It's like a magician's trick, but with science," the scientist chuckled. "Instead of coins, we measure the properties of pairs of entangled particles. And let me tell you, the more we measure, the more fascinating this phenomenon becomes. But, the key is to make sure that what we're seeing isn't just a fluke, you know, a stroke of luck. Hence, the science behind entanglement is equal parts amazing and rigorous."

The scientist paused, a mischievous grin on his face, "So, ready to flip some cosmic coins and test the limits of entanglement?

String Theory

This theory is taking the physics world by storm and for a good reason.

In the world of particle physics, things can get pretty small, like subatomic small. And that's where string theory comes into play. It suggests that the tiny building blocks of our universe aren't particles at all but tiny strings vibrating at different frequencies. Think of it like a tiny cosmic guitar string plucking away at the fabric of the universe.

But wait, there's more! These strings aren't just vibrating in our three-dimensional world; oh no, they're vibrating in a whopping ten or eleven dimensions! That's right; our universe may be even weirder than we thought.

The strings themselves exist in a higher-dimensional space, where the traditional laws of physics are but mere whispers. And, as if that weren't enough, these strings are believed to be the key to unifying all of nature's forces, including gravity, into one grand, harmonious theory.

String theory is still just a theory; it hasn't been proven or disproven. But it's got some pretty big names in physics talking, and it's definitely an exciting time to be alive if you're a science lover. Who knows, maybe one day we'll unlock the secrets of the universe and finally be able to answer the question, "What's really going on here?"

Chapter Three

Introducing Uncle Albert

Gather 'round, folks, and let me tell you about the ultimate physics prodigy: Albert Einstein! This dude was a theoretical physicist hailing from Germany who lived from 1879 to 1955, and let me tell you; he was the absolute boss of brains. His IQ was so high it could probably make a rocket go to the moon and back! His superior intellect allowed him to achieve amazing intellectual feats and conjure up revolutionary theories that made him one of history's greatest and most influential physicists and philosophers. In 1905, Einstein had a year so miraculous it's known as his "miracle year" (or, as the fancy people say, "annus mirabilis"). In that single year, he published four papers that rocked the very foundations of modern physics. He explained all sorts of cool ideas like the Photoelectric effect, Brownian Motion, and Mass-energy Equivalence. But the one that really tickles our fancy today is Special Relativity. Einstein opened our eyes to a mind-bending concept: space and time are connected in ways nobody had ever thought of before! But before we get too deep, let's talk about the man himself. Einstein was a bit of an introvert. He loved being left alone with his thoughts and always pondered how to solve theoretical problems. He was a quiet fella who just wanted to understand how the world works.

E = Energy, m = mass of object, c = the speed of light, 2 = squared

In 1940, he became a U.S. citizen and praised the American system for encouraging creativity. He was determined to unravel the mysteries of the universe by creating underlying principles and logical structures, and boy, did he succeed! Let's get into the juicy stuff. You know how Galileo proved that objects fall at the same speed no matter their mass? Well, Newton's theory said that twice the mass means twice the pull and twice the speed upon impact. But Einstein? He blew all that out of the water! Our man Albert came up with the idea that big objects bend and warp the fabric of space and time. And that's how gravity works, people! Can you even fathom that? Einstein completely flipped physics on its head with one single idea! He was the ultimate game-changer.

The Young Einstein

Let's hop on a time machine and go back to the days when Einstein was just a young lad, chasing after light beams like a puppy chasing after its tail, cooking up crazy ideas that would rock the world of physics to its core! And boy, did he deliver! He showed us that space and time are like taffy, stretching and squeezing to fit the situation like a pair of spandex leggings. But how did he prove this mind-bending concept? With a series of epic thought experiments, of course! And one of the most epic involved a lightning storm and a train track.

Hold up a Picture

Get ready to tap into your imagination! You hold in front of you a picture. The image is of a train track, with a train chugging along from left to right pulling a passenger carriage behind. And who do we have riding on this train journey? Why, a lone passenger gazing out the window, of course!

Now, let's add a little more excitement to the drawing. You are holding the drawing in front of you, gazing at the passing train. And just to make things even more interesting, there are two trees in the picture, tree number 1 on the left and tree number 2 on the right, both standing tall at exactly 1 mile away from you, the observer.

So there you have it, a visually stunning scene with a train, a passenger, two trees, one on each side, and you, the observer, looking on.

The train is chugging along at close to the speed of light

Now, imagine the scene coming alive and is no longer a picture but actually happening in real-time.

And, with a crash of thunder, two lightning bolts strike the two trees at precisely the exact moment, one mile away from you, standing and watching, in the foreground.

To you, the two lightning strikes happen simultaneously, with the light from both reaching your eyes at the exact same time. But things look a little different for the passenger on the train, who's speeding along at nearly the speed of light. To them, one lightning strike hits the tree ahead, and only after that does the other bolt hit the tree behind.

Why the difference? It all comes down to the train's motion. The passenger is moving toward the tree on the right, so the light from the bolt ahead reaches their eyes faster than the bolt striking the tree behind.

And what did Einstein deduce from all this? That events that appear simultaneous to one observer might not be happening simultaneously for another. And thus, the concept of simultaneity was forever altered! And, oh yeah, he also confirmed that the speed of light is constant anywhere in the universe, but time moves differently for objects at rest and those in motion.

Whoa, hold on! Have you ever heard of time playing a wild game of hide-and-seek? Well, it does! Depending on your velocity and proximity to massive objects like planets or black holes, it speeds up or slows down. The faster you travel, the more time decides to slow down!

Double Slit Experiment

Physics has this crazy experiment called the Double Slit Experiment that will blow your mind with its seemingly outrageous behavior of light beams! It's like watching a magic show, where light beams can be both waves AND particles and observing it affects its performance. It's like, 'What just happened?' But don't worry; I'll make it easy for you, the experiment showcases the unpredictably probabilistic nature of quantum mechanics. Or not even that! Just enjoy the magic show!

Imagine a laser beam that passes through two tiny slits and, whoa, splits into two waves! The waves go their separate ways, only to come back together and create a pattern of bright and dark bands on a screen behind the slits. If light were just a classic particle, this wouldn't happen. But it gets even crazier! Scientists have found that light is both a particle AND a wave! How is this possible, you ask? Welcome to the wild and wacky world of quantum mechanics and its notorious wave-particle duality!

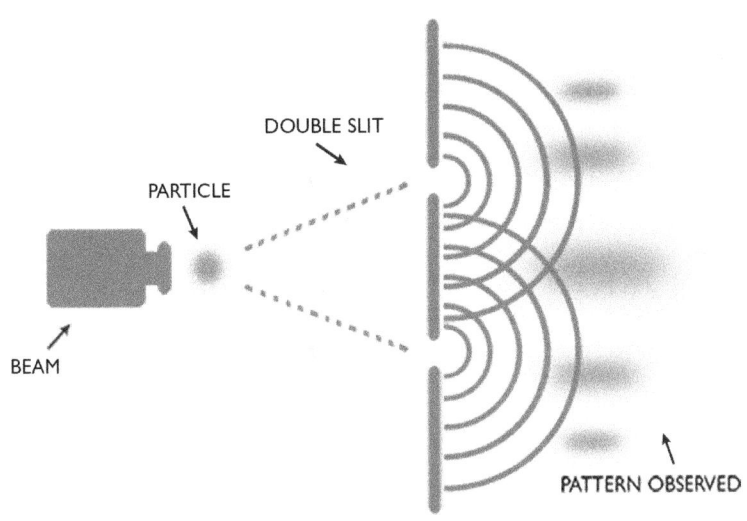

And get this; the observer has a role to play too. When you measure a quantum system, you're actually changing it. That's right, just by observing, you're influencing the outcome of the experiment. How's that for power?

The Double Slit Experiment has become a classic demonstration of the fundamental limitation of the observer's ability to predict experimental results.

Chapter Four

In The Loop - The Quantum Realm

The Loop Theory

This theory is like a magician ready to make all your quantum gravitational dreams come true. It's the marriage of two of the most powerful forces in physics - quantum mechanics and general relativity.

Picture this - instead of smooth, continuous space-time, loop quantum gravity thinks of it as a series of tiny, interconnected loops. These loops are the building blocks of the universe's fabric, and determine the fate of all matter within it.

This theory also predicts that the smallest size space can have is not zero but a tiny, non-zero value called the Planck length.

The Planck Length

The Planck length is the little guy that packs a big punch. This minuscule measurement is considered to be the smallest length that actually has any physical significance. And who do we have to thank for this incredible discovery? None other than Mr. Max Planck, the German physicist who was a genius ahead of his time and a pioneer in the world of quantum theory.

But why is the Planck length so darn important, you ask? Well, buckle up because this is where things get really interesting. It's considered to be the smallest length at which the laws of classical physics just can't cut it anymore, and quantum physics steps in to save the day. That's right, it's a fundamental constant of nature and plays a crucial role in the study of black holes, quantum gravity, and all the other exciting stuff that keeps physicists up at night.

This is like discovering that the universe's building blocks are made of LEGOs - they might be small, but they pack a big punch!

The Planck length is significant because it is considered to be the smallest length at which the laws of classical physics break down, and the principles of quantum physics become dominant.

This theory suggests that gravity is not a force but arises from the geometry of these space-time quanta. It's a mind-bending concept, but scientists are working hard to unravel the mysteries of loop quantum gravity and unravel the fabric of the universe! Who knows what other cosmic secrets will be uncovered as we delve deeper into this magical realm?

Have you ever heard the phrase "the devil is in the details"? Well, when it comes to understanding the fundamental fabric of the universe, we've got more than a few details to iron out! But we're scientists and philosophers, so we're not afraid of a little challenge.

Let's take a look at your computer screen, for instance. It might look like the images and text have clean, sharp edges. But, if you zoom in further, you'll see that what you thought were smooth edges are actually made up of tiny, individual pixels.

It's the same with space-time. If we could zoom in, we might find that time doesn't move smoothly into the future, but instead stumbles along, one pixel at a time.

Now, when it comes to theories about the universe's basic fabric, they're like leaves on a windy day - here one minute, gone the next. But that's okay because the journey is where the fun is! But that's just how the game is played. The more we delve into the nitty-gritty details, the more we seem to be uncovering new questions and mysteries.

The world of physics is a loony, insane place filled with big ideas about how our universe works. But if you want to play with the big kids, you've gotta bring your math game! That's right; for any theory about the fundamental laws of the universe to be taken seriously, solid mathematical proof needs to be backed up.

Think of it like a recipe for a super cool science cake. Sure, you can have all the best ingredients, but if you don't measure them out properly and mix them in the right order, your cake is gonna be a mess! Similarly, theories about the universe need to be proven with the right mathematical formulas, so we can be sure that they hold up to the rigors of scientific scrutiny.

So don't be fooled by all the dazzling ideas and colorful diagrams; at the end of the day, physics is all about numbers.

Quantum Physics

Now, we're about to delve into the mind-bending world of quantum physics! And you know what they say, "If it were easy, everyone would be doing it!" Well, not everyone, because according to the one and only Richard Feynman, the world-famous American theoretical physicist, he famously quotes, "I think I can safely say that nobody really understands quantum physics." This statement highlights the complexity and counter-intuitive nature of quantum mechanics, which even the greatest minds in physics have struggled to fully grasp. Yes, you heard that right; nobody really understands quantum physics. It's like trying to understand a magician's tricks, they seem so simple and straightforward, but in reality, it's a complex web of illusions. But hey, that's what makes the journey of discovery so much more exciting, right?

We have got two heavy hitters in the ring, ready to rumble! In one corner, we've got Quantum Mechanics, with its mind-bending rules and mysterious ways. And in the other corner, we've got Particle Physics, duking it out with subatomic particles and cosmic forces. These two giants of physics are ready to take on all comers and leave you in awe with their thrilling discoveries and mind-boggling concepts. So grab your lab coat, and let's dive into the action!

Quantum Mechanics

Quantum mechanics, also known as Quantum Physics, is like a magician's trick that blows your mind! It's the branch of physics that explains the behavior of particles on a really, really tiny scale. It's the study of everything small, from the tiniest particles that make up our universe to the things that make up those particles. We're talking about subatomic particles, the little critters like electrons, protons, and neutrons. If you thought regular physics was crazy, wait until you get a load of Quantum Mechanics, it's like a mini-universe all its own! The funny thing is, these little guys don't follow the same rules as the big things we see around us every day.

For example, according to classical physics, particles can be in only one place at a time. But in the quantum world, particles can be in two places at once, a phenomenon known as superposition. And when you try to observe a quantum particle, it immediately pops into one place or another, a process known as wave function collapse. It's like magic!

Quantum mechanics also introduces the idea of entanglement, where particles can be connected so that what happens to one instantly affects the other, no matter how far apart they are. As our old friend, Albert, said, "Spooky, right"?

So, if you thought physics was just about calculating force and mass, think again! Quantum mechanics is a whole new level of weird and wonderful that's still not fully understood, even by the experts. Just to remind you, it was Richard Feynman, who famously said, "I think I can safely say that nobody really understands quantum physics." Richard Feynman, the rockstar physicist, took the world by storm with his charismatic teaching style and killer lecture skills, in the middle part of the last century. This curious and adventurous mind had a passion for physics that couldn't be contained, and his many contributions to the field have made him a true icon of 20th-century science. From his ground-breaking work in quantum mechanics to his popular books and lectures, Feynman left his mark on the world and inspired a new generation of physics enthusiasts.

Particle Physics

Particle Physics is like a miniaturized superhero adventure where scientists are on a quest to save the universe - one particle at a time! It's all about uncovering the secrets of the tiniest building blocks of matter and energy and understanding the forces that bring them together. Our heroes, the particle physicists, use state-of-the-art equipment like particle accelerators and detectors to study the properties of subatomic particles, testing predictions and making new discoveries. They're on the lookout for heavy hitters, also known as "elementary particles," who are the stars of the show and include quirky characters like quarks - up quarks and down quarks, gluons, bosons, mesons, and even mysterious anyons. These particles are so small that they can only be detected through experiments. Still, the insights gained from studying them have massive implications for our understanding of the universe and influencing fields like quantum chemistry, quantum optics, quantum computing, super-conducting magnets, light-emitting diodes, and medical and research imaging, such as magnetic resonance and electron microscopy, cosmology, astrophysics, and technology.

Particle Physics is like a detective story where scientists are on a mission to uncover the secrets of the tiniest building blocks of our universe!

The science of Particle physics uses its powers for good in every aspect of life! From developing life-saving medical isotopes to creating superconductors that will change the world as we know it, particle physics is always ready to save the day. And that's not all - it's also got our backs in national security, industry, computing, and more. With particle physics on the case, there's no problem too big or too small that it can't handle!

The Star Experiment

Quantum mechanics is like the wild, untamed animal of physics - it's got some seriously mind-boggling behaviors! Remember, we talked about "spooky action at a distance." It's the idea that particles can "know" about each other instantly, even if they're light-years apart. Now, if you think about it, that's a major violation of the cosmic speed limit, which states that nothing can travel faster than the speed of light.

But hang on to your helmet, it gets even crazier! Scientists want to ensure there's no loophole in the experiments or hidden physical mechanism that creates the illusion of entanglement. So, a group of researchers decided to call in the big guns - they used starlight to test the theory. So, these scientists in Vienna, Austria, set up shop on the rooftops of their university and the Austrian National Bank with a couple of telescopes. They were on a mission to collect photons - the building blocks of light - with a specific wavelength.

And let me tell you, these bright stars were no joke - they were sending photons at 'em like a firehose! But the researchers were ready with their speedy detectors to register these cosmic photons' arrival in a mere subnanosecond.

These scientists measured 100,000 pairs of entangled photons, and what do you know? They found that these particles were truly entangled! And get this, the most distant stars used in the experiment were a staggering 600 light-years away! That means these photons had been emitting their cosmic message for 600 years!

But here's the real kicker, if these entangled photons were connected to distant stars, that connection had to have been established 600 years ago. As one professor of physics at MIT put it, "In order for some crazy mechanism to simulate quantum mechanics in our experiment, that mechanism had to have been in place 600 years ago to plan for our doing the experiment here today and to have sent photons of just the right messages to end up reproducing the results of quantum mechanics. So, it's very far-fetched."

So, what do the results tell us? More support for this "spooky" phenomenon.

Chapter Five

Very Special Relativity

You remember our old uncle Albert Einstein, the man who turned physics upside down with his Special and General Relativity theories and a genius who knew how to play with time and space?

In 1905, he took the world by storm with his book on Special Relativity, which explained how speed affects mass, time, and space. And then a decade later, in 1915, he upped his game with General Relativity, where he added gravity to the mix.

But who needs an introduction to the iconic equation that sums up Special Relativity? E = mc squared, baby! It's the equation that shows us how small amounts of mass can be converted into massive amounts of energy. And as we approach the speed of light, the mass of an object becomes infinite, making it impossible to go faster than light. Talk about a cosmic speed limit!

Special Relativity is all about ultra-high speeds and astronomical distances, making it perfect for discussions on huge energies. And General Relativity? Well, it inspired new realms of physics and science fiction, as we dream of traveling vast distances through time and space. Thanks, Uncle Einstein!

The Flat Earth

So, there are still people in this world of ours, who think the Earth is flat. That's like thinking the sky is green and the grass is blue. But, to be fair, if you're standing on the Earth, it does appear flat. But if you get a little altitude, flying in an airplane or if like you were in space, you'd see the curved edge of the Earth in all its spherical glory.

Foucault's Pendulum

Now, for the flat-Earth folks, there's a fun device that'll change their minds.

It's not just a simple experiment - it's an elegant one too! This French physicist, Léon Foucault, sure knew how to make science stylish back in 1851 when he first showed it off to the world.

Picture it: a long, heavy pendulum suspended from the ceiling high above, swinging freely in every direction. And as the Earth turns beneath it, the plane of the swing starts to rotate, oh so slowly. It's like a dance, a clockwise or counterclockwise waltz, depending on the hemisphere, that is so captivating to watch.

And what's causing this pendulum dance, you ask? Why, it's none other than the Earth's rotation! As the Earth turns, the pendulum swings back and forth, but the ground beneath it moves, causing the plane of the pendulum's swing to rotate.

And get this, folks, the time it takes for the pendulum's plane of oscillation to do a full spin is the same length as a day! That's right, 24 hours of pure spinning goodness.

Foucault's pendulum made waves for showing us the Earth's rotation in a way that was groundbreaking at the time. But even today, it's still a simple and effective way to explain physics to people of all ages. Science museums around the world keep it on display, and it's still a popular attraction.

But believe it or not, people used to think the Earth was flat. I'm talking about way back in the days of Aristotle, over 2,000 years ago. It wasn't until he saw the shadow of the curved Earth on the Moon that he proposed the Earth was round. Fast forward a couple of thousand years to the time of Columbus, and most folks still believed the Earth was flat. They thought they were at the center of the universe and that everything revolved around them!

Copernicus

Enter, Copernicus, about 500 years ago. He had the audacity to suggest that Earth wasn't the center of the universe but instead revolved around the sun. The church was shaken and accused him of heresy, but we scientists knew he was onto something - we didn't want to get locked up or executed for agreeing. Nowadays, we've got these fancy words like geocentric and heliocentric to describe whether we're center stage or playing second fiddle to the sun. But it wasn't until Kepler and Galileo busted onto the scene in the early 1600s that we really started to get the hang of this solar system thing. They thought all the planets were held in orbit by magnetism and that gravity kept us from floating off into space.

Sir Isaac Newton

Sir Isaac Newton, the mad scientist of the 17th century, was a man ahead of his time! This genius was not only a mathematician but also a physicist, astronomer, and philosopher. He left his mark on history as one of the most influential scientists to ever exist. But it's not just objects that are affected; time is too! According to Einstein, time passes slower in stronger gravitational fields and faster in weaker ones. This is known as time dilation, folks. And all of this happens in a four-dimensional fabric, where space and time are combined into one intense ride.

But what exactly did Sir Isaac do to earn this title? Well, for starters, he rocked the scientific world with his laws of motion and universal gravitation, which basically gave us an understanding of how the world moves. He also developed the reflecting telescope, allowing us to see the stars in a new light. His groundbreaking work, "Mathematical Principles of Natural Philosophy," laid the foundation for classical mechanics and provided a comprehensive explanation of the natural world.

Now, let's talk about Isaac's Three Laws of Motion. They are simple enough for a child to understand but powerful enough to change the world.

Things don't change unless a force makes them change. This is the idea of inertia.

If you want to move an object, you need to use force. The amount of force needed depends on the mass of the object and how fast you want to move it.

Every action has an equal and opposite reaction. This law has been a favorite of kids everywhere because it's just an excuse to push your friends and family around!

Sir Isaac Newton, a legendary scientist, came up with a theory that rocked the world... literally! He claimed that his laws of motion and gravitation applied to everything in the universe and created a brand new math - calculus - to back it up!

Back in the day, this was like a blast of fresh air in a stuffy room. But, let's be real, who wants to read heavy scientific theories? So, Sir Isaac made it simple with a thought experiment.

Imagine yourself on top of a mountain with a big ol' cannon and a cannonball. And, just for the sake of this experiment, let's pretend there's no air resistance. You fire the cannonball and... boom! It flies for a bit and then falls back to earth, thanks to gravity pulling it down. But, what if you add more gunpowder to the mix? The cannonball flies further and further before it falls back to earth. And what if, just what if, you fire the cannonball with so much gunpowder that it escapes the pull of Earth's gravity and keeps orbiting around the Earth, that's what we call orbital velocity.

But, if there was no gravity, the cannonball would fly off in a straight line forever. That's what Sir Isaac Newton's theories are all about! Minds blown, folks?

It's all about the math, baby! That's right, rocket science is powered by equations and calculations that put the "aim" in "aim for the stars". Every launch is a carefully orchestrated dance of numbers, ensuring that the rocket and its precious payload are propelled to the correct speed and trajectory to reach the heavens.

And once that capsule is in orbit, it's more number-crunching to keep it steady. The International Space Station circles the Earth like a math-savvy ballerina, holding a precise position with the help of critical speeds and expert calculations. No tumbling down to Earth or shooting off into space for this space station. It's all thanks to the power of mathematical precision!

Newton's theory of gravity demonstrates that an object's mass is the critical factor, not the object's size. A small thing can have a more gravitational effect than a larger object when it has more density. A mathematical equation expresses that bodies attract each other with force proportional to mass and inversely proportional to the square of the distance between them.

Let me explain the difference between mass and weight.

Weight is different from mass. The force of gravity gives us the weight of an object. The mass of an object will never change, but based on its location, the weight of an item can vary.

Gravity, it's not just a force, it's a matter of MASS appeal! The bigger the mass, the bigger the pull - that's the idea behind Newton's law of attraction. And let's not confuse mass with weight, folks. If you weigh 100 pounds on Earth, you would be a weightless wonder in space. Weight can change depending on where you are in the universe, but mass stays the same - just like the love for pizza, it's constant!

The Apollo 15 astronauts proved that gravity works the same on the moon as it does on Earth. They dropped a hammer and a feather simultaneously on the moon and guess what? They hit the surface at the exact same time! Mind blown, right?

The secret behind it all is that space and time get warped by massive objects, like the Earth and the Moon, creating a 'fabric of space' that pulls on everything and creates valleys in that fabric.

Think of the Moon as a ball rolling down a slope in the fabric of space - and that slope was created by none other than the massive Earth! Our Moon is falling towards the Earth, pulled by its gravitational pull like a moth to a flame.

Who knew space could be so full of drama!"

Picture this, a blazing hot rock hurtling through space, minding its own business when suddenly, BAM! It gets smacked with Earth's gravitational pull and crashes into our planet in a fiery explosion. But wait, what if it had just the right amount of momentum? It would avoid the fiery fate and instead be captured in a delicate dance around our planet, soaring in a beautiful orbital tango.

And who knows, perhaps our lovely Moon was once just a wandering celestial body, snatched up by Earth's gravity and trapped in its orbit. Or, the wild theory goes, it was once a part of Earth itself before a massive collision with a space object caused it to be hurled out into the great unknown.

Either way, the mysteries of our universe never cease to amaze!

Have you ever bounced a ball on a moving train and felt like time stood still? It's the coolest magic trick around! The ball bounces back and forth from your hand to the floor, never deviating from its path - or so it seems. But if you had a bird's eye view of the train, you'd see that ball moving down the aisle like a boss, defying all laws of physics! That's what we call relative movement, my friend.

And speaking of relative, have you ever noticed how time and space are always together like two peas in a pod? When you make plans with a pal, you always pick a date and a location. That's because time and space aren't two separate things - they're spacetime! And if you think time is constant, you're in for a shock.

Thanks to Einstein's theory of general relativity, we now know that time isn't so straightforward. It's all about speed and gravity. The faster you jet off in your spaceship, the slower time ticks for you compared to your friends back on Earth. By the time you come back, they'll be all wrinkly and gray, but you'll still have your youthful glow!

And if you think that's mind-bending, wait until you hear this - the further away you are from a massive gravitational object, the faster time flies for you! There's no such thing as absolute time, so don't even try to keep up!

So, I'm gonna sum it up like this…

Spacetime! It's like a big, stretchy trampoline where everything happens, and all physical objects exist. But wait, it's not just any ordinary trampoline. No, no, no. It's a trampoline with a mind of its own, shaped and molded by the presence of matter and energy.

So, here is the man again, the myth, the legend: Albert Einstein and his theory of general relativity. He's the mastermind behind this crazy idea that spacetime can curve, just like a bowling ball rolling along on that same trampoline. And with this curvature comes the origin of gravity, causing objects in the universe to be drawn toward massive objects like planets and stars.

But it's not just objects that are affected; time is too! According to Einstein, time passes slower in stronger gravitational fields and faster in weaker ones. This is known as time dilation, folks. And all of this happens in a four-dimensional fabric, where space and time are combined into one intense ride.

Chapter Six

The Terrifying Black Void - the Black Hole

Just the name sends shivers down your spine, doesn't it? It sounds like the universe's version of a bottomless pit - a black void that sucks in everything in its path – a gravitational force so intense; it's enough to make your hair stand on end (if you had any in space, that is).

But don't be fooled; black holes are far from empty! They're actually formed by a super intense gravitational force created when a dead star collapses and draws everything around it, in. And we're talking a lotta mass, baby - anywhere from 100 to 100,000 times the mass of our sun!

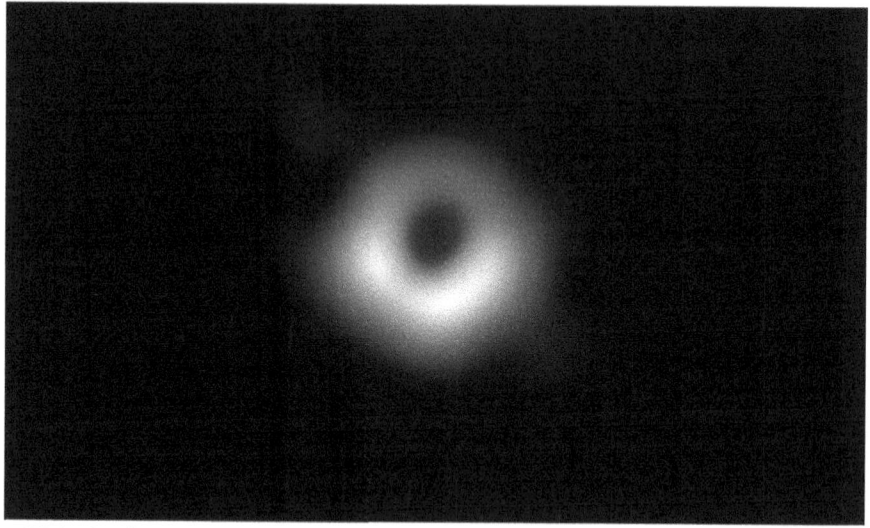

This is the supermassive Black Hole at the center of our own Milky Way Galaxy, called Sagittarius A, It is 4 million times the size of our Sun

These cosmic monsters are so dense that they warp the spacetime around them. Scientists still have no idea what's going on inside them, but according to the laws of physics, if you find yourself on the brink of being sucked in by its gravity near what's called the Event Horizon, time would slow down so much that you might never enter the black hole! That's right, time comes to a complete standstill. And, if you were to survive and make it back to Earth, you'd be younger than everyone else. Now, how's about that for a time-traveling adventure?

We know that light travels at a breakneck speed of 186,000 miles per SECOND, which is so fast, it's almost impossible to wrap our minds around it. But here's the catch - everything with mass can't travel at that speed, not even close. The bigger the mass, the harder it is to get moving and the more energy it takes. And let's be real, reaching the speed of light is a pipe dream unless you're a beam of light or a radio wave with no mass.

These days, GPS – Global Positioning Satellites, are a testament to the impact gravity has on time. Each one of these satellites has an atomic clock that ticks with a precision of 1 nanosecond or one billionth of a second. But the clocks on the ground tick a bit slower because of Earth's gravity, so the clocks on the satellites have to be adjusted to keep the global navigation system functioning smoothly.

Chapter Seven

Surfing the Gravity Wave

Gravity waves, or as I like to call them, space-time ripples, are the latest buzz in astrophysics. These waves are like little tremors in the very fabric of the universe, caused by massive objects moving around, such as black holes or exploding supernovae. Einstein predicted them in 1916, but it took until 2015 for us humans to finally detect them with our fancy gadgets.

Now, let me tell you, these waves are no joke. They're so weak you could sneeze and create a bigger disturbance. But don't let their feeble nature fool you; they carry an insane amount of energy across the cosmos.

It's like this - imagine you're sitting in a pool, and suddenly, someone decides to do a cannonball right next to you. The water ripples and the waves reach you, causing you to bob up and down. That's exactly what happens when a gravity wave passes through Earth, except the pool is the entire universe, and the cannonball is two black holes colliding billions of light-years away.

Speaking of which, the first gravity wave detected by LIGO, our trusty gravity wave detector, was caused by the collision of two black holes. (LIGO stands for Laser Interferometer Gravitational-Wave Observatory). These weren't just any black holes; mind you, they were each about 30 times the mass of the sun. That's like two superheavyweights in a cosmic boxing match, duking it out in slow motion.

Since then, LIGO has been working overtime, detecting dozens of gravity waves from all sorts of cosmic events, from merging two neutron stars to the effects of a supermassive black hole at the center of a distant galaxy. It's like the universe is having a rave, and we're finally able to hear the music.

Detecting these waves has opened up a whole new way of observing the universe, giving us a glimpse into the secrets of gravity and space-time. It's like a whole new dimension of science has been unlocked, and we're just getting started.

So there you have it, folks - space-time ripples, the ultimate cosmic tremors that are shaking up the world of astrophysics. It's like Christmas morning, but instead of presents, we get mind-blowing discoveries about the universe. Who knows what else we'll discover with these bad boys? Exciting times, my friends, exciting times.

In conclusion, time, space, and gravity are like three musketeers, always sticking together. What we observe and measure depends on the observer, but they're all equally accurate. Gravity affects time, and the more an object speeds up, the heavier it gets.

Chapter Eight

Chasing Unicorns: The Quest for a Unified Theory

Imagine a grand cosmic party where all the different branches of physics are hanging out, doing their own thing. You've got gravity in one corner, electromagnetism chatting it up with particles in another, and the strong and weak nuclear forces doing their own thing. Suddenly, in walks the Unified Theory, and everyone stops and stares. So there you have it. The Unified Theory is like the ultimate party planner of the universe, bringing all the different branches of physics together and showing them a good time. It's like the cool kid in high school that everyone wants to hang out with, except instead of parties and social status, it's all about understanding the fundamental nature of the universe. And who knows, maybe one day we'll all be dancing to the same cosmic tune.

Gravity, the old grandpa of the group, mutters something about how he's been around for billions of years and hasn't seen anything like this before. Electromagnetism starts fiddling with her hair nervously, wondering if she's wearing the right outfit. And the strong and weak nuclear forces just stand there, looking stoic and unimpressed.

But the Unified Theory doesn't care. It struts over to the center of the room, slaps down a set of equations, and starts explaining how it's going to bring everyone together. "You see," it says, "all of you are just different aspects of the same thing. You're all part of the grand cosmic dance, and if we can just get you to play nice with each other, we'll have a beautiful symphony of the universe."

At first, the other branches of physics are skeptical. They've never really gotten along before, and they're not sure they want to start now. But the Unified Theory is persuasive. It points out how gravity and electromagnetism are just two sides of the same coin and how particles and nuclear forces are all intertwined. It even throws in some fancy math to back up its claims.

Slowly but surely, the other branches of physics start to come around. They start brainstorming, coming up with new ideas and ways of looking at the universe. They realize that the Unified Theory might be onto something and that maybe it's time to put aside their differences and work together.

And before you know it, the cosmic party is in full swing. Gravity and electromagnetism are dancing together, the particles are high-fiving the nuclear forces, and everyone is having a grand old time. The Unified Theory is beaming with pride, knowing that it was the one that brought everyone together.

Chapter Nine

Breaking the Space-Time Continuum: Understanding Time Travel

The ultimate time-bending concept that's got us all on the edge of our seats for centuries. The idea of traveling through time to experience history or the future is mind-blowing. But honestly, can we really do it? The science world is debating it, and we're here to explore the nitty-gritty of time travel and the potential effects it could have on our bodies.

So, what exactly is time travel? It's the act of traveling through time, either to the past or the future. This concept has been the stuff of legends and popular culture for eons, and it's become a topic of scientific inquiry in recent years. Basically, time travel is a way of moving through time beyond the normal flow of time we experience in our everyday lives.

Now, the million-dollar question: can we really time travel? While some think it's mere fiction, others argue it could be possible in theory. The idea of time travel is based on the concept that time is not a fixed and immutable thing but rather a fluid concept that can be manipulated in certain circumstances. And theories like Einstein's theory of relativity or the concept of wormholes suggest that time travel could be a possibility.

Assuming time travel is possible, how could we do it? Well, that depends on the specific method of time travel being used. First up, we have the classic "time machine" option, made famous by the likes of H.G. Wells and Doc Brown. Hop in your souped-up DeLorean or creaky old Victorian device, crank some levers, dials, and whoosh! Off you go, hurtling backward or forwards through time at breakneck speeds.

Mental time travel is the ability to mentally project oneself backward or forward in time to relive past experiences or imagine future ones. It's like having a personal time machine in your brain!

When you mentally time travel, you can vividly recall past events, people, and places as if you were there again. You can even imagine yourself in the future, living out your dreams and accomplishing your goals.

But mental time travel isn't just a fun party trick. It's actually an important cognitive ability that helps us learn from our past mistakes, plan for the future, and navigate the complexities of the present.

Imagine you're about to take an important exam. By mentally time-traveling to your past experiences, you can recall similar tests you've taken in the past, and use that information to help you study better and improve your performance.

Or, let's say you're trying to make a big decision about your future career. By mentally time-traveling to the future, you can imagine yourself in different career paths and see how each one plays out, helping you make an informed decision.

So, the next time you catch yourself daydreaming about the past or the future, don't worry, you're not going crazy. You're just engaging in a little mental time travel, which can be a valuable tool for personal growth and success.

If we're talking about a time machine, we just enter it, set the desired date and time, and in a split second, we're there. But if we're going through wormholes, things get trickier. The process of entering and exiting a wormhole would be like walking on thin ice - it could be extremely dangerous and even fatal.

Let's talk Dr. Who! The show follows the adventures of a time-traveling alien known as the Doctor, who travels through time and space in his TARDIS. The TARDIS is a unique time machine that's disguised as an old-fashioned police box, which used to be common in England, and it's able to travel through time and space by using a technology known as time vortex manipulation. Sounds fictional, right? But some scientists think it may be possible to manipulate the fabric of space-time to create a wormhole.

So, what are the potential mind-boggling effects of time travel? If we're traveling through a time machine, we're good to go. But if we're using wormholes or other means of manipulating space-time, we might experience some severe effects. For example, time travel could mess with our aging process, affecting our physical and mental health. Imagine traveling

100 years into the future - you'd be skipping over 100 years of aging, which could significantly affect your body.

And what psychological impact does time travel have on a person? Imagine being flung into a completely different time period with different norms, customs, and values. You'd probably feel like a fish out of water, and that's just the start! You might even experience some serious culture shock that could lead to feelings of disorientation, alienation, and even trauma. It could be hella frightening!

And what about this? Time travel could also mess with a person's sense of identity. Imagine if you went back in time and met your ancestors, only to realize they were nothing as you imagined. Suddenly, everything you thought you knew about your family history and your place in the world would be turned upside down. It's enough to make your head spin!

Then there is the concept of paradoxes!

The "grandfather paradox" is a real mind-bender! Here's the deal: let's say you hop in a time machine and head back to a time before your grandpa was hitched and had a kid. Now, if you do something crazy like knock off your grandpa or stop him from getting hitched to your grandma, then your own parents would never have been born. And if your parents never existed, that means you wouldn't exist either! Talk about a major dilemma! It's like a chicken-and-egg situation but way more complicated. So, basically, if you go back in time and mess with your family tree, you could end up disappearing into thin air. That's pretty wild, right?

The grandfather paradox is just one challenge to the idea of time travel and the concept of causality. If time travel were possible, would altering events in the past change the future and create a paradox? Only time will tell. But for now, let's keep our fingers crossed for a TARDIS or a DeLorean to come to our rescue!

Chapter Ten

The Multiverse Theory: Are We One of Many or Many of One?

The Multiverse is one strange and wonderful concept that's sure to blow your mind! I mean, just think about it - there could be an infinite number of universes out there, each with its own unique properties and laws of physics. It's like a cosmic choose-your-own-adventure book!   But at the end of the day, the multiverse theory is just one more reminder of how bizarre and wonderful the universe truly is. We may never know for sure whether there are other universes out there, but the fact that we're even contemplating the possibility is pretty mind-blowing. Who knows - maybe someday we'll find a way to peek into these other universes and see what they're all about. Until then, let's just sit back, relax, and enjoy the ride. Who knows what kind of craziness the universe has in store for us next?

So, what exactly is the multiverse theory? Well, it proposes that multiple universes exist alongside our own, each with unique properties and characteristics. And we're not just talking about slight variations here - we're talking about universes where the laws of physics are completely different from our own. I'm talking about universes where gravity doesn't exist, where time flows backward, and where unicorns are real (okay, maybe not that last one, but you get the idea).

One of the most popular models of the multiverse theory is the Many Worlds Interpretation. According to this model, every time a quantum measurement is made, the universe splits into multiple universes where every possible outcome is realized. So, for example, if you flip a coin and it lands on heads, there's a universe where it landed on tails instead. And in that universe, you probably didn't get that promotion you were hoping for. Sorry, pal!

Another model of the multiverse theory is the Bubble Universe Theory. This theory suggests that our universe is just one of many "bubbles" that exist in a larger multiverse. Each bubble has its own physical properties and laws of physics. This means there could be a bubble universe out there where the laws of gravity are entirely different from ours or where the speed of light is much slower. Can you even imagine what it would be like to live in a universe where you could outrun light? You'd be like the Flash, only cooler.

Now, I know what you're thinking - this all sounds like some crazy sci-fi mumbo-jumbo. But here's the thing - the multiverse theory is a genuine scientific theory that's gaining traction in string theory. According to this theory, the universe is composed of tiny strings that vibrate in more than four dimensions. That's right, more than four dimensions! We're not in Kansas anymore, Toto.

So, what are the implications of the multiverse theory? Well, for starters, it challenges our understanding of reality and our place in it. The idea that our universe is not unique suggests that there could be an infinite number of other universes, each with unique forms of life. I mean, think about it - there could be a universe where everyone has tentacles instead of arms. Or a universe where dogs are the rulers of the world (let's be real, though, they're already ruling our hearts, aren't they?).

The multiverse theory also raises some pretty trippy philosophical questions about the nature of reality and the role of observation in shaping the universe. According to the theory, the universe is not a fixed entity but a collection of possibilities. Our observations and actions may determine which possibilities become reality. This raises the question of whether reality exists independently of our observation of it, or whether our observations shape the nature of reality itself. It's like that old philosophical question - if a tree falls in the forest and no one is around to hear it, does it make a sound? Instead of a tree, it's a universe, and instead of sound, it's... well, I don't know what it is. But you get the idea.

Of course, as with any scientific theory, the multiverse is not without its skeptics. Some argue that the theory is untestable and unscientific, while others claim that explaining certain observations in physics and cosmology is necessary. But you know what? I say - why not embrace the wackiness of it all? I mean, life is already pretty bizarre as it is. We're just a bunch of primates living on a rock hurtling through space at breakneck speeds. If the multiverse theory is true, then at least it makes things a little more interesting.

But let's be real - the idea of an infinite number of universes can be pretty overwhelming. I mean, where do you even begin to wrap your head around that? It's like trying to count all the grains of sand on a beach. It's just not possible. So, let's focus on the practical implications of the multiverse theory.

One of the most exciting implications of the multiverse theory is that it could help us solve some of the biggest mysteries in physics and cosmology. For example, the theory could explain why the laws of physics in our universe seem so finely tuned for life. If there are an infinite number of universes, then it's possible that we just happen to live in one where the physical constants are just right for life to exist. It's like playing a game of darts - if you throw enough darts at the board, eventually, one is going to hit the bullseye.

The multiverse theory could also help us understand why the universe is expanding at an accelerating rate. According to the theory, the expansion of our universe could be influenced by the gravitational pull of other universes. In other words, our universe is not expanding into empty space but rather into a space that is filled with other universes. This could explain why the expansion rate of our universe seems to be increasing over time.

But let's embrace the fun implications of the multiverse theory. If there are an infinite number of universes out there, then it's possible that there are universes where we're all superheroes, or where we're all giant insects, or where we're all just really good at math. The possibilities are endless.

Of course, the multiverse theory is not without its challenges. One of the biggest challenges is that it's difficult to test. After all, if there are other universes out there, then they are by definition, beyond our reach. We can't observe them directly or interact with them in any meaningful way. This makes it difficult to come up with experiments that could prove or disprove the theory.

Another challenge is that the theory raises some pretty thorny philosophical questions. For example, if there are an infinite number of universes, then is there really such a thing as free will? Or are all of our choices predetermined by the laws of physics in our universe? It's a tough question and one that's sure to keep philosophers busy for a long time.

But at the end of the day, the multiverse theory is just one more reminder of how bizarre and wonderful the universe truly is. We may never know for sure whether there are other universes out there, but the fact that we're even contemplating the possibility is pretty mind-blowing. Who knows - maybe someday we'll find a way to peek into these other universes and see what they're all about. Until then, let's just sit back, relax, and enjoy the ride. Who knows what kind of craziness the universe has in store for us next?

THE EXPANDING MIND

"Look up at the stars and not down at your feet. Try to make sense of what you see, and wonder about what makes the universe exist. Be curious." – Stephen Hawking

Before we get to the different types of stars out there, let's take a moment to breathe… After all, the notion that our universe is always expanding is pretty mind-blowing!

It's crazy to think that it gets bigger and bigger, unlimited by boundaries and not needing any space to expand into… But when you really think about it, it's not such an unfamiliar concept – your knowledge is like that.

You're learning reams as you get further through this book, and it doesn't mean you have to get rid of existing knowledge to make room for it. The human brain has an unlimited capacity for learning – just as the universe has an unlimited capacity for expansion.

Better yet, we can spread knowledge, taking it even further and expanding it out to reach other people… and you can do that right now without setting foot out your front door.

By leaving a review of this book on Amazon, you'll show new readers where they can find a wealth of information about the universe – presented in a fun and engaging way that it won't hurt their brains to process!

Simply by letting other readers know about your experience of reading this book and what they'll find inside it, you'll show them where they can find a ton of information to expand their knowledge.

Thank you for your support. The universe is such an exciting thing to be a part of – and I think we all deserve to know more about it.

>>>Please be so kind as to leave a review using this link...

amzn.to/45j5ICY

SECTION TWO

Galaxies, Constellations, Black Holes, Dark Matter

Chapter Eleven

Starry Nights: Exploring the Wonders of the Night Sky

When you stand outside at night, the weather is clear, the sky is exceptionally dark and assuming you have excellent vision, look up. You will see stars that stretch from horizon to horizon. You are looking at around 2,500 stars. Now, there are a couple of galaxies that we can actually see with the naked eye - The Andromeda Galaxy and The Triangulum Galaxy. But they're super far away, at around 17 quintillion miles. I mean, who has time for that kind of long-distance relationship?

Now, counting the number of stars in the sky is like counting the number of ants at a picnic - it's near impossible! But we'll give it a shot.

The stars in the night sky are like those party animals that never seem to have a curfew - some nights, they're out in full force, creating a brilliant light show, and other nights they're nowhere to be found.

The number of stars you can actually see depends on a few things - where you're hanging out, the time of year, and whether or not the light pollution patrol (i.e. city lights) is crashing the party. If you find yourself in the middle of nowhere, with a clear view of the sky and no city lights in sight, you might be able to spot a jaw-dropping 5,000 or so stars with your naked eye! But, if you're stuck in the heart of a brightly lit city, the stars might be playing hard to get, and you might only be able to see a measly few hundred.

And then, of course, there is the moon - it's like that overbearing relative who always seems to show up uninvited. When the moon is full, it shines so bright that it steals the spotlight from all the other stars, making it difficult to see anything but the biggest and brightest ones. But, on a moonless night, the stars come out to play, and you might be able to spot up to 10,000 of them!

So, the bottom line is that counting the stars in the night sky is a bit like a game of hide-and-seek - it all depends on the conditions!

Sure! The night sky is a constantly changing and dynamic phenomenon that has captivated humans for thousands of years. The stars, planets, and other celestial objects visible in the night sky are incredibly distant from Earth and their light has taken thousands or even millions of years to reach us. The moon, which is Earth's closest celestial neighbor, has a significant impact on the night sky. Just as an aside, get this - the sun's rays have taken 8 minutes to travel to the moon. The light from the moon is that sunlight reflecting back to us on earth and it takes one and a third seconds for it to reach our eyes, which shows you how mighty close the moon is to the earth. Actually, it is only 240,000 miles away.

The moon's phases change throughout the month, creating a familiar sight for many people and playing a role in human history, including using lunar calendars and navigation at sea.

Constellations, patterns of stars that form recognizable shapes, have been used for centuries to navigate and tell stories and legends. Some of the most well-known constellations include Orion, the Big Dipper, and Cassiopeia. The Milky Way, a barred spiral galaxy, is also visible in the night sky and can be seen as a hazy band of light stretching across the sky on clear, dark nights.

In addition to being a natural wonder, observing the night sky can also be a scientific pursuit. Astronomers study the night sky to learn about the structure and evolution of the universe, the formation of stars and planets, and the search for extraterrestrial life. Advances in technology have allowed astronomers to make incredible discoveries, from the detection of exoplanets to the observation of distant galaxies.

Overall, the night sky is a reminder of the vastness of space and our place within it, and its study has been integral to the advancement of human knowledge.

Are you ready to wrap your mind around just how big this place really is? Let me tell you; it's bigger than you can imagine. In fact, if you lined up every atom in your body, they wouldn't even reach halfway across the Milky Way galaxy. Can you believe that?

And traveling across that galaxy in a speedster airplane at 500 mph? Well, buckle up because you'll be on that ride for 100,000 years! But that's just a hop, skip, and a jump compared to crossing the vastness of space.

Our Milky Way, one of thousands and thousands, takes us 250 million years to cross from one end to another. And our entire solar system, a mere speck in the grand scheme of things, is just one tiny dot. But you know what makes us so special? We can gaze upon the universe with our own eyes and witness the stars twinkling above us. So, let's take a moment to be thankful for this incredible perspective.

The heavens are filled with all sorts of stars, each one shining bright and unique in its own way. Stars make up most of the visible mass of the Universe.

Let's take a look at some of the most interesting stars out there!

First up, we have the **red giant**s. These stars are the divas of the universe, living life to the fullest before finally burning out in a dazzling supernova. They're like your fun-loving uncle who never grows up, always making the party a little brighter.

Next, we have **white dwarfs**. These retiring stars have run out of fuel and are biding their time until the final curtain call. They may be small, but they pack a punch and have a surprisingly long lifespan. Think of them like your wise old grandparent, always there to give you advice and make you smile.

And then there are the **neutron stars**, the pint-sized powerhouse of the universe. These stars pack more mass than the sun into a ball just a few miles across!

Last but not least, we have **black holes**. These enigmatic objects are the mysterious strangers of the universe, drawing everything in with their powerful gravity. They may be invisible, but they're impossible to ignore. Think of them as the enigmatic loner in a dark corner at the party, drawing all eyes their way without saying a word.

So, have you ever looked up at the sky and thought, "Wow, those stars are all different shapes and sizes. Some are like little peas in a pod, and others are like giant basketballs up there!" Well, let me tell you, those stars are really just like people - some are hot-headed and fiery, while others are just chillin' at a cool few thousand degrees.

Then we have the drama queens of the star world - when they start running out of fuel, they go out with a bang and explode into a supernova. Talk about a diva move! But others prefer a more understated exit and implode into a tightly packed star like a white dwarf or a black hole. You know, gotta keep it classy.

Now, I know it sounds like there are a bajillion stars out there, but we've actually estimated it to be around 70 billion trillion. That's, like, more than all the cats on the internet combined! And get this - stars recycle. When one star dies, it forms new stars like some cosmic recycling program.

Our Sun is the star of our neighborhood show, powering everything in our solar system. And let me tell you, those planets are high-maintenance divas too. But what's really cool is that stars tend to cluster together like a giant celestial clique to form star clusters and galaxies. It's like Mean Girls in Space!

And don't even get me started on how big and bright those stars are. They make our Sun look like a tiny little firefly. And if you're lucky enough to see them on a clear, dark night, you can see out 19 quadrillion miles.

Chapter Twelve

From Bright Giants to Dim Dwarfs: Understanding the Star Magnitude Scale

Back in the day, the Greeks had a pretty impressive trick up their togas - they created a way to measure the brightness of stars to create a catalog of celestial wonders. The scale was first introduced by the ancient Greek astronomer Hipparchus and is logarithmic in nature. And you know what? That scale is still in use today but with some serious upgrades!

Imagine a horizontal line with a big, bold zero in the center. To the right of zero, we've got our fainter objects, marked off with plus 1 to plus 30 increments. And to the left of zero, we've got our bright and shining stars, marked off with minus 1 to minus 30 increments. Each step on this scale represents the brightness of the object in the sky in terms of its apparent magnitude.

Here's the deal: the further to the left you go, the brighter the objects get. And the further to the right, the fainter they get. For example, the full moon is shining like a star and is placed at minus 12.7, while the Sun, the brightest of them all, is placed at a whopping minus 27. Sirius, the brightest star in the sky, is at minus 2, and even Polaris, the North Star, is only at plus 3.

The naked eye can only see up to plus 7 magnitude, but with a little help from a telescope, you can gaze at objects with a magnitude of minus 7 and minus 14. And get this: each increment is two and a half times brighter or dimmer than the step next to it. The planet Venus is a fourth-magnitude stunner, shining at minus 4.

Our Sun is yellowish and falls in the middle of the visible part of the color spectrum. The greater the size of the star, the hotter and bluer it is. The less massive the star is, its color becomes redder.

So there you have it, the Star Magnitude Scale, a measure of the brightness of the stars that continues to light up our skies and imaginations.

Chapter Thirteen

Connecting the Dots: Exploring Constellations

Constellations are like a cosmic choose-your-own-adventure book. They're patterns in the sky made up of stars that can be interpreted as anything from animals to mythical figures to household objects. Before the trusty compass came along, sailors relied on these star formations to navigate the high seas. And guess what? The North Star, also known as Polaris, was their trusty sidekick. It stays put in the Northern hemisphere, residing in the constellation of Ursa Minor - also affectionately known as The Plough. Those sailors were hardcore! Can you imagine being lost on the high seas and using the stars to find your way home? These celestial beauties are like a group of jet-setters, always on the move and changing positions with the seasons. But don't worry, they're not flaky like that one friend who always cancels plans - they keep their distances from Earth so vast that any perceived changes are just a cosmic illusion. Can you imagine, our ancestors who were just starting to walk upright and ex-

CONNECTING THE DOTS: EXPLORING CONSTELLATIONS

ploring the world, would have seen the same constellations shining above them, like a cosmic connect-the-dots puzzle? It's pretty mind-boggling to think about the history behind these stars that have been guiding us for generations. And let's not forget their practical uses! These twinkling dots have been our cosmic calendar for centuries, marking the passing of time and helping us navigate through the night sky. So, next time you're gazing up at the stars, take a moment to appreciate these dazzling travelers and the stories they have to tell.

Not only did constellations help with navigation, but they also served as an ancient calendar. Different constellations would appear in different seasons, marking the passing of time. But you don't have to worry about the stars moving around and messing up your plans - they're so far away that any changes are virtually undetectable by the naked eye. What our ancient ancestors saw, when they gazed into the heavens, living 30,000 years before us, is pretty much the same as we see today.

There are 88 named constellations in the sky, which the International Astronomical Union recognizes:

Here are a few classic constellations to look out for:

Orion: This constellation is a real crowd-pleaser, shaped like the legendary Greek hunter. His belt is three stars in a row, and the sword hanging off it even has a faint nebula - talk about a weapon with special effects!

Cassiopeia: Named after a Greek queen, Cassiopeia resembles a "W" in the sky. Not only was she royalty, but she had a killer sense of direction - she's located near the North Star!

North Star: Also known as Polaris is located more or less directly above Earth's north pole along our planet's rotational axis - it's called the north celestial pole. As our planet rotates through the night, the stars around the pole appear to rotate around the sky. If you were looking at a time exposure photo of the night sky, you would see a circle around the celestial pole. The farther a star is from the pole, the larger the circle it travels. Because the North Star is so close to the celestial pole, it traces out a very small circle over 24 hours, so it always stays in roughly the same place in the sky, and therefore it's a reliable way to find the direction of north.

Ursa Major: Also known as the "Great Bear," Ursa Major contains the Big Dipper asterism. And if you want to find the North Star, just look for the Big Dipper! This group of seven stars forms a shape that's easy to spot.

Scorpius: This constellation is shaped like a scorpion and is the real deal - it actually looks like one! The red star Antares marks the heart of the scorpion, adding a little flair to this cosmic critter.

These celestial beauties are like a group of jet-setters, always on the move and changing positions with the seasons. But they're not flaky like that one friend who always cancels plans - they keep their distances from Earth so vast that any perceived changes are just a cosmic illusion. Can you imagine, our ancestors who were just starting to walk upright and exploring the world, would have seen the same constellations shining above them, like a cosmic connect-the-dots puzzle? It's pretty mind-boggling to think about the history behind these stars that have been guiding us for generations. And let's not forget their practical uses! These twinkling dots have been our cosmic calendar for centuries, marking the passing of time and helping us navigate through the night sky. So, next time you're gazing up at the stars, take a moment to appreciate these dazzling travelers and the stories they have to tell.

Chapter Fourteen

The Expanding Universe

Picture the universe as a giant cosmic bubble, expanding at an incredible speed. That's right, the universe is growing, and it's doing it fast! In fact, it's expanding so quickly that it's mind-boggling. The current consensus is that the Universe's expansion rate is 163,000 miles per hour and accelerating based on the data brought to us by the Hubble telescope.

Think about it, every single thing in the universe - from stars and galaxies to black holes and mysterious dark matter - is getting farther apart from each other. It's like a cosmic game of "red light, green light," with the universe constantly saying "go!" And get this, the expansion is accelerating! That's right, the universe is not only expanding, but it's expanding faster and faster.

But don't worry, the universe is so vast that you won't notice the expansion with your own two eyes. The change is slow and gradual, but it's happening. In a way, it's like the universe is taking a deep breath, inhaling and expanding in all directions.

The Greeks sure had an eye for the stars! Just like the first humans, they saw the same celestial wonders in the sky that we see today. And you know what they did with that? They brought the stars to life with mythical stories and folklore. These tales were about heroes, beasts, and everything in between, all depicted through the patterns of the stars.

The Zodiac

Take Aquarius for example. This constellation is one of the 12 signs of the zodiac, sandwiched between Capricornus and Pisces. The Greeks saw a young man named Ganymede in the stars. Legend has it that Zeus fell in love with him and brought him to Mount Olympus to be the cupbearer to the gods. Ganymede was granted eternal youth, and the constellation is said to resemble a picture of the young man, complete with a pot of liquid pouring out of it.

The stars have their own personalities too. A star's magnitude determines its brightness, and some stars even have proper names based on their brightness and the constellation they call home. For example, Carina, Hydra, Orion, Columba, Centaurus, and Ursa Minor are all stars with proper names.

And just when you thought the stars couldn't get any cooler, each area of the sky is associated with one of the 88 officially recognized constellations. These boundaries were agreed upon by the international community in 1930. So, whether a star has a proper name or not, it'll always have a name based on its brightness and the constellation it belongs to. It's like a celestial club where every star has a place!

Next, I'm gonna show you the different types of stars that are out there, shining their little hearts out. Get ready to learn about the stars that light up our nights and keep us company on those lonely starry nights.

There are about 10 broad categories of stars, with dozens of sub-categories! So, get ready to have your mind blown by the diversity of these celestial wonders. Are you ready? Let's go!

Chapter Fifteen

A Journey into the Heart of Stars

Stars come in all shapes, sizes, and colors. Some are hot-headed and fiery, while others are cool and calm like cucumber.

First up, we have the blue giants. These guys are the hotshots of the star world, radiating a fierce blue light and burning themselves out relatively quickly. They're like the celebrities of the universe - always burning bright, but usually crashing and burning just as fast.

Next, we have the red dwarfs. These tiny stars may not be as flashy as their blue counterparts, but they make up for it in longevity. They can burn for billions of years, which is great for any civilizations that happen to be living on planets orbiting them.

Then there are the yellow stars, like our very own sun. These stars are just right - not too hot, not too cold, but juuust right for planets like ours to orbit them and support life. They're like the Goldilocks of the star world.

And let's not forget about the white dwarfs, the burnt-out husks of stars that have already lived their best life. They're like retirees of the universe, still shining but not as brightly as they used to.

Main Sequence Stars

Main Sequence Stars that are stable are known as the Main Sequence stars that light up the night sky. Unstable stars are named Non-Main Sequence stars.

The Main Sequence stars are the bad boys - the all-star players of the galaxy, making up a whopping 90% of all the stars out there.

Our own Sun is a Main Sequence star, a G-type yellow dwarf, and it's just one of the many sizes these stars can come in. From tiny little one-tenths of the mass of our Sun to a massive 200 times bigger, Main Sequence stars are the real MVPs of the sky. Think of them as super athletes, fueled by hydrogen in their cores, with the ability to transform it into helium atoms.

But how do these stars come to be, you ask? Well, it all starts with a cloud of dust and gas. Gravity pulls it all together, creating a dense star known as a protostar. But if it's not quite big enough, it'll be stuck in protostar purgatory, forever known as a brown dwarf. But if it becomes a massive body of gas and dust, then BOOM, the star collapses into itself and becomes a Main Sequence star, ready to light up the night with its incredible power.

And when the temperature within the star reaches a scorching 10 million degrees Kelvin, it's game on, baby! The nuclear reaction starts and a brand new star is born.

The Variable Stars

These youngens are just hitting the scene, with only around 10 million birthdays under their belt. But don't let their youth fool you, these stars have got some serious personality!

See, they're still trying to figure out who they are and what they want to be when they grow up. One day they're shining bright, lighting up the night sky, and the next day they're feeling a little down and their brightness fluctuates. It's like they're going through a rebellious phase, and we all know how that goes, right?

These stars are still growing, gobbling up gas and dust from nearby nebulae like they're at an all-you-can-eat buffet. The strong solar winds from all that material falling onto the star's surface create sunspots and an accretion disc, which might even turn into planets someday!

But don't worry, these stars will eventually settle down and become Main Sequence Stars. It takes a few million years for the combination of gravity pulling inward and nuclear fusion pushing outward to balance out and stabilize the star. Our own Sun took a leisurely 50 million years to become the shining star it is today. So, these Variable Stars have a long and exciting journey ahead.

A massive Main Sequence star, 10 times the mass of our Sun, can burn for 20 million years, while our Sun can survive for only 10 million years.

Red Dwarf Stars

A Red Dwarf star is about half the mass of our Sun and can survive for 80 to 100 billion years. It's interesting to note that the Universe is a mere 13.8 billion years old. This fact suggests that planets orbiting these suns are more stable for a long time, and the chances are that life forms will be given a better opportunity to appear and evolve.

Double Stars

Many stars attract each other and are known as double stars that orbit each other at various distances. 60% of stars are in these binary systems, and other systems comprise 4 stars orbiting each other in a delicate ballet.

So, if you thought our Sun was a long-lived party animal, think again! A Main Sequence star that's a real heavyweight, 10 times the mass of our old pal the Sun, only lasts for a wild 20 million years. Meanwhile, the Sun is just living it up for 10 million years. But don't worry, there's still hope for a long life. Enter the Red Dwarf star, a little lighter in the mass department, around half the size of our Sun, but they can keep the party going for 80 to 100 billion years! That's pretty impressive considering the entire universe is just a teenager at 13.8 billion years old. So, you know what that means? The planets orbiting these Red Dwarfs have a better chance of sticking around for a while and potentially creating a stable environment for life to flourish. Some stars can't stay away from each other. 60% of stars are part of binary systems, and there's even a group of four stars that perform a mesmerizing dance as they orbit each other. Cosmic relationships at their finest!

Brown Dwarf Stars

A Brown Dwarf is a failed star whose mass wasn't enough to ignite a nuclear fusion reaction

Brown Dwarfs are like the underdogs of the celestial world - they never quite made the cut to be a full-blown star. They're the celestial equivalent of the guy who always strikes out at the plate, but hey, they still have a place in the universe!

So, like, the basic idea is that stars are burning hydrogen, which is pretty hot stuff. But eventually, they run out of hydrogen and things start to get a little wild. The core of the star starts contracting, while the outer part starts expanding, and boom - we've got ourselves a Red Giant. It's like the star version of a midlife crisis. It gets all big and bloated, and it's not as bright as it used to be. Sad, really. Eventually, the star's envelope becomes loosey-goosey and starts to drift away, until all that's left is the core, which is now a White Dwarf. And then, poof! It's gone like last night's pizza. But get this - our very own Sun is gonna become a Red Giant in about 5 billion years! And when that happens, things are gonna get crazy. Like, seriously cray. The Sun is gonna expand so much that it'll engulf Mercury, Venus, and probably Earth. That's right, folks - we'll all be toast. And just to put it into perspective, a Red Giant would be so big that it could fill up the entire space between the Sun and Jupiter. Mind – Blown?

Supernovas

Much bigger-sized stars than our Sun die in a more dramatic way. A star with 10 times the mass of the Sun will burn through its material in 100 million years and collapse into an ultra-dense white dwarf. It ultimately explodes in a burst of brilliant light called a Supernova that can outshine an entire galaxy. The energy it will radiate will be more than the energy used during the Sun's lifetime.

Get this, ladies and gents! A star is simply a ball of hydrogen that's burning bright. But as it runs out of fuel, the core starts to contract under gravity, while its outer layers puff up, transforming into a Red Giant. This fusion-powered beast expands many times its original size, but it's also a lot dimmer than before. The end of a star's life has arrived!

The outer layer starts to peel off, leaving only the core, now known as a White Dwarf, to slowly fade away. As I mentioned before, in around 5 billion years, our Sun will transform into a Red Giant, swallowing up Mercury, Venus, and possibly even Earth. Just imagine, the Red Giant will be so big it could fit Jupiter's orbit inside of it!

But for stars bigger than our Sun, the death is much more explosive.

From a blazing Supernova to a slow and steady decay, stars have many different ways to say goodbye. But what a show they put on!

Stars can die in different ways, from a Supernova explosion to slow, steady decay.

Blue Giant Stars

The largest stars in the Universe are called Blue Giant Stars - Stars don't get more massive or hotter than blue giant stars. Their color appears blue, their surface blazing at about 20,000 degrees Kelvin.

So, let me explain here the term degrees Kelvin. First of all, there is a temperature measurement called Absolute Zero. This is the lowest temperature theoretically possible in the Universe and is equivalent to minus 459 degrees Fahrenheit.

The dictionary definition of Kelvin is "the Standard International base unit of thermodynamic, first introduced as the unit used in the Kelvin scale". (It's a tough one to understand.) Scientists use the Kelvin scale because they want a temperature scale reflecting thermal energy's complete absence. Lower temperatures mean fewer vibrations in the material and no vibrations at Absolute Zero. Less heat equals fewer vibrations.

Our Sun shines at 6,000 degrees Kelvin. If you remember the brightness of celestial objects scale, Blue Giants are in the minus 5 or 6 range.

The Blue Giant Stars are the largest and hottest stars in the whole dang Universe! With temperatures reaching a sizzling 20,000 degrees Kelvin, these blue beacons of light are seriously hot to the touch. And when I say hot, I mean hotter than a jalapeño pepper on a hot summer day.

Degrees Kelvin

But what's a Kelvin, you ask? It is named after a British physicist, William Thomson, also known as First Baron Kelvin – don't ask me why! Think of Kelvin as a special temperature measurement just for scientists – the name was used in his honor. It's a scale that starts from absolute zero - the coldest temperature possible in the whole universe. And when I say cold, I mean colder than a penguin's toes on a winter day.

See, scientists like to use the Kelvin scale because it takes into account thermal energy and the absence of it. The lower the temperature, the fewer the vibrations in a material. And at absolute zero, there are no vibrations at all!

Now, back to the Blue Giants. With a surface temperature of 20,000 Kelvin, they are the brightest stars out there. Our sun, in comparison, shines at a mere 6,000 Kelvin. And on the brightness scale of celestial objects, Blue Giants are in the minus 5 or 6 range - which basically means they are brighter than a room full of 100-watt light bulbs!

Meet Rigel, the rockstar of the Universe! This blue giant, named after the brilliant and stylish Nigel, is a true giant among stars. With a size that's a whopping 25 times that of our humble Sun, this supergiant is a true force of nature. And, boy, does it know how to turn up the heat! Rigel's surface temperature is sizzling at a scorching 50,000 degrees Kelvin. But, with such high temperatures comes great responsibility. This star is burning through its fuel at a breakneck pace, so hold onto your hats, because in a few hundred million years, Rigel will put on an explosive show that will rival the biggest fireworks display you've ever seen. But, don't worry, our Sun will still be chugging along for another 12 billion years.

Neutron Stars

Picture this, folks - giant stars at the end of their lives, collapsing under their own gravity and leaving behind either Black Holes or Neutron Stars. It's like a cosmic game of "now you see me, now you don't." And let me tell you, the Neutron Stars are the wild children of the space party! They're like a giant star that went through an epic breakup, smashed into a ball the size of a city, and given a gravitational pull 2 billion times stronger than Earth's. And get this, they're spinning faster than a vinyl record on a turntable, up to 43,000 times a minute! They're like the red pepper flakes on top of your pizza, adding an extra kick of excitement to the universe.

Some of these Neutron Stars are pulsars and magnetars. They're like the popular kids in high school, but with superpowers. Pulsars have jets of material ejected at nearly the speed of light, which can be seen from Earth as a flashing beam. And magnetars? They have magnetic fields more powerful than anything else known to us in the Universe! They're like that one kid in class who always had the coolest gadgets.

Scientists call these flashy stars pulsars because they emit X-ray beams that pulse like a disco ball. These beams are so narrow that you need a radio telescope to catch them. And get this: the beam only shows up if it's pointing directly at Earth but these high-maintenance stars can be useful, too.

They spin at different speeds, which makes them handy for space travel. You can use them to pinpoint your spacecraft's location, which is a real lifesaver when you're lost in the cosmic void.

Now, let's talk about magnetars. These stars are like the Kardashians of the universe. They have an insane magnetic field that's stronger than anything else out there. It's so powerful that it can even mess with the shapes of atoms! That's some serious bling, if you ask me.

But here's the catch: magnetars burn out fast. They only last for about 10,000 to 100,000 years, and then they fizzle out like a bad reality TV show.

Now, let's talk about Millisecond Pulsars, aka "Black Widow" pulsars. They steal material and energy from their neighboring companion stars, becoming faster and more powerful. It's like they're the vampires of space, but instead of blood, they need energy.

But don't worry, it's not all doom and gloom. In fact, all the heavy metals on Earth, like gold and platinum, are believed to have originated from these cosmic collisions. That's right, folks, we're all made of "colliding-star stuff." So, next time you're wearing your favorite necklace or ring, just remember - it's not just a fashion statement, it's a piece of the universe itself.

The Terrifying Black Holes

Ah, black holes! These space cowboys love to party and an accretion disc of superheated gases and dust surrounds them. But black holes are far from empty! They eat and eat until they burst into X-ray outbursts that can be measured. The Massachusetts Institute of Technology even found eight black hole binaries where a star was orbiting them. What a wild party! These guys have a strong gravitational pull that not even light can escape. Imagine having a bouncer that intense at your party! Black holes are just like cosmic pranksters who suck in anything that gets too close. If you're not careful, they'll play the ultimate "No Re-Entry" policy game with you. But fear not, you can observe these party monsters from afar. They will make stars orbit around a central point, like they're dancing to the beat of their own cosmic music. And when they are in the mood to show off, they'll blast out high-energy radiation, screaming to the universe, "Hey, look at me!" There are different types of black holes for different occasions. There are the stellar black holes, which are like the lightweights of the black hole world. Then we have intermediate black holes, the middleweights, and finally, we have the supermassive black holes, the heavyweights of the black hole world. These guys are so massive they can have masses equivalent to billions of suns. That's like having the biggest person in the room, but a billion times bigger! But it's not just their size that's impressive, it's what's at the center of these bad boys that's really mind-blowing. Scientists have speculated for years about what's at the heart of a black hole, but we still don't know. Some even think there are parallel universes, wormholes, and time travel in there! It's like the ultimate treasure hunt in the universe.

So, how fast do stars orbit a black hole? It's like asking how fast a cheetah can run or how quickly your stomach rumbles after a big burrito. The answer depends on a few factors.

First off, we gotta talk about the black hole's mass. You know how some folks can't resist that extra slice of pizza? Well, black holes are kind of like that. The more matter they gobble up, the bigger they get, and the stronger their gravitational pull becomes. So, if you've got a hefty black hole on your hands, stars will be zipping around it faster than you can say "warp speed."

The distance between the star and the black hole also plays a role. It's like those carnival rides where the closer you sit to the center, the more intense the spinning becomes. If a star is cozying up to a black hole, it's gonna feel the full force of that gravity and start orbiting like a bat out of hell. On the other hand, if it's keeping a safe distance, it can take a leisurely stroll around the black hole without breaking a sweat. For example, stars in close proximity to the supermassive black hole at the center of the Milky Way galaxy, known as Sagittarius A*, orbit at speeds of up to incredible speeds of seventeen million miles per hour. However, stars that are farther away from the black hole will orbit at slower speeds.

The black hole has an insatiable hunger. It's like that one friend who always has to order extra appetizers even though you're already stuffed. If a black hole starts chowing down on nearby stars, it can totally mess with their orbits. Suddenly, they're careening off in wild directions and making NASA scientists break out in a cold sweat.

The speed at which stars orbit a black hole is a complicated dance between the black hole's mass, the star's distance, and the black hole's appetite. It's like trying to predict which way a cat will dart when it sees a laser pointer - you can make some educated guesses, but you never quite know for sure.

Back in 2015, a group of smart cookies detected gravitational waves - you know, the ripples in space-time that Einstein predicted a century ago. It's like a spacey wave that happens when two black holes decide to have a little tango and merge together. We're talking about GW150914, a massive event that took place 1.3 billion years ago

The Laser Interferometer Gravitational-Wave Observatory, known as LIGO, is the ultimate cosmic detective!

LIGO is a real show-off, boasting not one, but two gigantic interferometers located in Hanford, Washington, and Livingston, Louisiana. These babies work together to detect the teeny tiny ripples in space-time caused by crazy events like black holes going boom or neutron stars going smash.

And let's talk about the size of these things! They're so long, they make Shaq look like a garden gnome. But, hey, size matters, especially when it comes to detecting the tiniest fluctuations in the universe.

But how does it all work? It's simple, really. They shoot laser beams down long tubes, bounce 'em back and forth, and measure any changes caused by gravitational waves passing through. Just like that, they're able to detect some of the most violent events happening millions of light-years away.

Since then, LIGO and other facilities have been observing black hole mergers like it's their day job - and they're pretty darn good at it!

Now, I know what you're thinking: "But how can we study something that doesn't emit light?" Well, hold your horses, because there's more to the story. Astronomers have been studying black holes for decades by checking out the light they emit. Sure, light can't escape a black hole's event horizon, but the tidal forces around it can heat up nearby matter to millions of degrees, causing it to emit radio waves and X-rays. And if that's not enough, some of the stuff orbiting closer to the event horizon can be hurled out, forming jets of particles moving at breakneck speeds that emit radio, X-rays, and gamma rays. These supermassive black hole jets can extend for hundreds of thousands of light-years into space - now that's a lot of spacey razzle-dazzle!

Black holes are given names. Some of them are just pure gold, like V616 Monocerotis, or "V-Mon." It's like the black hole equivalent of a stage name, and it's just begging to be turned into a punchline. Then there's the black hole with the catchy moniker, Cygnus X-1. It's like a rock star, with a name that just rolls off the tongue. And then there's the black hole in the galaxy NGC 4889, nicknamed "Sleeping Beauty." It's like a celestial fairy tale coming to life.

The stars in our own S Cluster are zipping around the black hole Sagittarius A like kids on a merry-go-round, with speeds reaching a dizzying 18 million miles an hour. These celestial bodies are caught in the black hole's gravitational grip, making them careen around its event horizon, in elliptical paths hurtling at breakneck speeds and getting thrown out billions of miles away in their orbital path. It's wild! Some stars complete the orbital trip in just four years, while others take a leisurely ten.

But it gets better! When massive stars go supernova, they blast out clouds of heavy elements that eventually form new planets. Scientists believe that black holes like Sagittarius A, located at the center of our galaxy, may have even played a role in shaping the formation of galaxies, including our own Milky Way. Meanwhile, our solar system is orbiting the center of the Milky Way at a speed of 514,000 miles per hour, and we complete a full orbit around the galaxy in a whopping 230 million years. Who knew a little black hole could be such a big deal?!

Our own Sun is a Main Sequence star, a G-type yellow dwarf

Chapter Sixteen

Unraveling the Mystery of Dark Matter and Dark Energy

And now, let's talk about the shadowy superstar of the universe, Dark Matter, the elusive and mysterious substance that has scientists scratching their heads and reaching for their telescopes! Despite making up an estimated 85% of the total mass in the universe, this cosmic phantom remains unseen, undetected, and downright maddening. So, dark matter remains one of the greatest puzzles of our universe, a cosmic conundrum that's both fascinating and frustrating. It's like trying to catch a mischievous leprechaun - you know it's there - you can't quite grab it! But scientists won't give up, oh no they won't! Solving the mystery of dark matter is one of the central goals of modern astrophysics and cosmology, and who knows, one day we may finally catch this cosmic enigma and put it under a microscope for the world to see.

It's like a cosmic game of "Where's Waldo?" except instead of a red and white-striped shirt, we're searching for a substance that's so slippery and evasive, it's as if it's playing hide-and-seek with us! But we can feel its presence, oh, yes we can! Scientists have observed the gravitational effects of dark matter on stars and galaxies, causing them to move in ways that can't be explained by visible matter alone.

Think of it this way, dark matter is the cosmic glue that holds the universe together, shaping galaxies, guiding stars, and leaving us wondering what on Earth (or beyond) it could possibly be made of! Some theories suggest that dark matter is composed of exotic particles, such as WIMPs or axions, while others propose it could be more down-to-earth particles, like neutrinos. But the truth is, no one really knows.

Matter is what you see and feel all the time. A teacup is made up of matter, a tree is made up of matter, and an elephant is made up of matter. Oxygen, hydrogen, and all of the gases are made up of matter even though many gases are invisible. But what is not made up of matter is the unseen part of the Universe and cannot, at this time, be measured. Only its effects are observable and measurable.

So, dark matter remains one of the greatest puzzles of our universe, a cosmic conundrum that's both fascinating and frustrating. It's like trying to catch a mischievous leprechaun - you know it's there - you can't quite grab it! But scientists won't give up, oh no they won't! Solving the mystery of dark matter is one of the central goals of modern astrophysics and cosmology, and who knows, one day we may finally catch this cosmic enigma and put it under a microscope for the world to see.

Enigmatic Dark Energy

Dark Energy, the mysterious force behind the accelerated expansion of our Universe, is unlike anything we've ever seen before. It has nothing to do with Dark Matter. the Universe has a bit of a curveball to throw our way. After the massive explosion that was the Big Bang, scientists were pretty confident that the Universe would slow down from all the excitement and eventually come to a halt. But, hold on to your lab coats, because around 8 billion years ago, the Universe started to speed up! You heard that right, it's expanding faster and faster, defying all the laws of physics that we thought we knew.

This little hiccup in our understanding of the Universe led to the creation of the term "Dark Energy". And before you start imagining a black, ominous force taking over the cosmos, let's just clear one thing up: "Dark" here doesn't mean its color, it means the mysterious, unknown factor. Dark energy, my friends, is one of the biggest mysteries in cosmology. It's believed to make up 68% of everything in the universe, with the matter we can actually see, "baryonic matter," only accounting for 5%. The rest is made up of the mysterious dark matter. But here's the real kicker, dark energy acts like Einstein's anti-gravity force, but its origin and nature remain a mystery. The real head-scratcher is why dark energy started dominating the universe's expansion rate billions of years after the Big Bang. If it's around now, why wasn't it there all along? Mysteries upon mysteries, folks!

Chapter Seventeen

Quantum Theory: When Particles Don't Play by the Rules

Ladies and Gentlemen, gather around because we're about to take a trip down the rabbit hole of the most bizarre and mind-bending theory in physics - Quantum Theory. We talked about it before, but it's worth repeating!

Imagine a world where particles don't play by the rules, where they can be in two places at the same time, and where they can communicate with each other instantaneously regardless of the distance between them. That, my friends, is the world of Quantum Mechanics.

Now, let's think about classical mechanics, the physics of the everyday world, where things have definite positions, definite speeds, and definite causes for those speeds. It's simple and intuitive, right? Well, hold onto your space helmets because we're about to throw that all out the window.

In quantum mechanics, particles can be described as both particles and waves at the same time. This duality is known as wave-particle duality and it still confounds scientists to this day.

Now, I know what you're thinking, "This is all well and good, but how does it actually work in the real world?" Well, let me give you a practical example. One of the most famous applications of quantum mechanics is quantum cryptography. This is a method of transmitting secret messages using quantum particles. The beauty of this method is that if someone tries to intercept the message, they will inevitably change the state of the particles, alerting the sender and receiver that the message has been compromised. It's like having two cats on opposite sides of the world, both dead, alive, and sleeping at the same time, and if someone tries to mess with one of the cats, the other cat instantly knows.

But, what's the catch? Well, the catch is that quantum mechanics is incredibly difficult to understand and even harder to explain. It's like trying to teach a cat quantum mechanics - it just won't get it. In fact, even the greatest minds in physics have struggled with the strange and unpredictable world of quantum mechanics.

Albert Einstein famously referred to it as "spooky action at a distance" and referred to quantum entanglement as "ghostly non-local connections". The father of quantum mechanics, Niels Bohr, once said, "If you aren't confused by quantum mechanics, you don't understand it."

So, in conclusion, quantum mechanics is a strange and bizarre world where particles don't play by the rules.

SECTION THREE

Our Solar System, Space Objects, and Exoplanets that exist Far and Beyond

Chapter Eighteen

Discovering the Diversity and Wonder of Solar Systems in the Universe

Solar systems are basically like the cool kids' clique in the universe. They consist of a star, which is the popular kid, and a group of celestial bodies that are always orbiting around it like the entourage of loyal friends.

It all starts with a big cloud of gas and dust, which is like the wild party that everyone wants to attend. As the party gets going, things start to heat up and the gas and dust condense to form a protostar, which is like the center of the dance floor where all the action is.

As the night goes on, the protostar gets bigger and bigger, attracting all sorts of attention from the other celestial bodies at the party. They start to orbit around the protostar like the cool kids flocking to the popular kid.

The whole thing can get pretty wild, with planets, moons, asteroids, and comets all jostling for position around the star. And just like in high school, the structure and composition of the solar system can vary widely depending on the personalities and conditions involved.

But the coolest thing is that we've discovered all sorts of solar systems beyond our own, from ones with multiple stars to those with giant planets that like to live dangerously close to their host star. Who knows what other crazy parties are going on out there in the universe? We are sure to find out someday soon.

A Family Portrait: Examining the Sun and Its Orbiting Planets

Picture this: Our Solar System, a cosmic dance party revolving around the dazzling star of the show, our sun! Despite the massive amount of material swirling around the Sun, when you consider the vast distances between everything, it's a mere blip in the vastness of space. So, what's the story behind our solar system's formation? How did it all come to be? Let's find out!

But let's go back in time, back to when our Sun was just a hot, dusty ball of gas and dust, eager to become the dazzling star it is today. It took a cool 50 million years to get its act together finally, but once it did, BOOM! It became a Main Sequence star, a rare G-type yellow dwarf, one of the 90% of stars in the universe that fuse hydrogen to form helium in their cores.

So, how did our Sun become the center of attention? It all started with a wild, swirling disk of material called a solar nebula, caused by the collapse of a stellar cloud of dust and gas, maybe even from the shockwave of a nearby supernova. As gravity took over, the material compressed into a dense star, a protostar. Our Sun was born, powered by hydrogen fusion and taunting all the material in its orbit with its relentless gravitational pull. The material, including planets, moons, asteroids, meteoroids, and other debris, all revolving around our sun, like guests at a wild party, all moving at a steady rate without any friction to slow them down.

And the party doesn't stop there!

Our Solar System is moving at a mind-blowing pace of 514,000 miles per hour, as it orbits the center of the Milky Way galaxy. It takes a full 230 million years for us to complete one orbit around the galaxy! So, next time you look up at the night sky and see the sun, planets, and other celestial bodies, just remember, they are all part of this incredible, never-ending dance party, and we are just lucky enough to be along for the ride!

Imagine our scorching hot star, the Sun, at the center, surrounded by a fleet of celestial bodies – there's Mercury, the speedy messenger; Venus, the sultry beauty; Earth, our home sweet home; Mars, the red rebel; Jupiter, the king of them all; Saturn, with its stunning rings; Uranus, the quirky one; and Neptune, the deep blue dweller. And then there's the dwarf planet, Pluto, plus countless moons, asteroids, comets, and meteoroids – it's a cosmic merry-go-round! But, let's not forget, it wasn't always this way.

Despite the massive amount of material swirling around the Sun, when you consider the vast distances between everything, it's a mere blip in the vastness of space. So, what's the story behind our solar system's formation? How did it all come to be? Let's find out!

The Coffee Bean Experiment

A legendary experiment took place aboard the International Space Station, where an astronaut got to play with some roasted coffee beans in zero gravity. Picture this - our trusty astronaut gently opens his hand and releases a bunch of coffee beans into the cabin, and they float around like mini asteroid-type rocks. But hold on there - what's happening? The beans slowly drift toward each other and startlingly form a cluster, clinging to each other like a space-bound community. Who knew coffee beans could be such physics masterminds?

This experiment just blew the lid off a massive truth bomb! Sir Isaac Newton himself revealed that an object's gravitational mass determines just how much gravity it's gonna throw around with the other objects in the vicinity. But it gets even crazier – this bad boy also measures how much gravity it's gonna suck up from the other objects in the area! That's right, every object out there is playing a cosmic game of attraction, with a force that's proportional to their combined mass and inversely proportional to how far apart they are.

Now, imagine this happening on a cosmic scale, where objects of varying masses and gravities come together, forming larger and larger bodies. With each new addition, the gravitational pull grows stronger, until you have massive celestial bodies like planets and stars. And that my friends, is how our solar system was born!

Recently, a team of researchers went on a mission to uncover the secrets of the past, and what they found was nothing short of mind-blowing. It turns out that the early days of our solar system were a lot more chaotic than we ever imagined.

Let's rewind the clocks to 4.6 billion years ago when cosmic gas and dust were swirling around the Sun, just waiting for the formation of Earth and the other planets. Sound familiar? That's right, It's just like the coffee beans in the zero gravity experiment; the dust particles started to stick together, forming rocks of all shapes and sizes. These rocks were moving and orbiting the Sun at breakneck speeds, creating a chaotic dance in the early solar system.

But the fun doesn't stop there; the researchers took a closer look at some fallen-from-heaven metallic asteroids and found that they were once so hot from the radioactive decay of isotopes that they were glowing! And get this; they were able to determine how quickly they cooled down by analyzing the asteroids. What they found was that the asteroids had crashed into each other with such force that they broke apart into smaller pieces and cooled down rapidly when exposed to the bone-chilling temperatures of space. This all happened just a mere 8 to 12 million years after the solar system's formation. And if you think that's wild, wait until you see the computer simulations of the solar system's development during that time. It was a total cosmic circus!

After the Sun was formed, it was surrounded by a cloud of gas and dust known as the solar nebula. This nebula slowed down the asteroids just like air resistance slows down a speeding car. But, the solar winds and radiation eventually blew the nebula away, and the asteroids were left to collide and careen into each other like crazy bumper cars.

Can you imagine the chaos? It was like a cosmic demolition derby, with rocks the size of houses, mountains, and even bigger crashing into each other with such force that they were shattered into pieces and sent flying in unpredictable directions. Some of the pieces burned up as they hurtled towards the Sun, while others were blasted off into the vast emptiness of interstellar space.

And all of these asteroids became the building blocks for the formation of our very own planet! So, the next time you step outside, take a moment to thank those crazy asteroids for helping create the world we call home.

There are different hypotheses as to how the planets were formed.

A team of researchers took a closer look at 18 different iron meteorites and found some juicy information about the asteroids they once belonged to.

The team dissolved the meteorite samples to isolate elements like palladium, silver, and platinum and then used a trusty mass spectrometer to measure the abundance of isotopes of these elements. But what are isotopes you ask? Think of them as identical twins that share the same protons but have a different number of neutrons.

The researchers found that these metallic asteroids were heated up by radioactive decay in the first few million years of the solar system. As they cooled down, a specific silver isotope produced by radioactive decay began to accumulate. By measuring the present-day silver isotope ratios, the researchers could determine when and how quickly the asteroids cooled down.

And the results are in! The cooling was rapid and likely caused by severe collisions with other objects. This resulted in the insulating rocky mantle being stripped off and the metal cores exposed to the cold of space. The timing of these collisions was even more precisely dated with the help of Platinum isotope measurements.

All the asteroids had collisions almost simultaneously, within a timeframe of 7.8 to 11.7 million years after the formation of the solar system, indicating that this period must have been a very chaotic phase of the solar system.

The team combined their results with the latest computer simulations of solar system development to come up with a theory for why this chaotic phase occurred. The culprit? The dissipation of the solar nebula, the gas that was left over from the cosmic cloud, gave birth to the Sun. For a few million years, the solar nebula orbited the young Sun, slowing down the objects in its path like air resistance slows a moving car. But once the nebula disappeared, the asteroids accelerated and collided with each other like turbo-charged bumper cars.

Our Backyard Planets

Planets are the most intricate and diverse objects in the universe. Planets that orbit around our Sun might be regarded as just the result of scrap material randomly jostled around and forming a clump of material, eventually becoming a planet. Our planets in our solar system demonstrate the wide variety of worlds that have been created with much complexity in astronomical, geologic, chemical, and biological processes.

Have you ever looked up at the night sky and wondered how those twinkling balls of rock, gas, and magic became what they are today? The planets! The fascinating objects in the cosmos, and here we are, lucky enough to have a few in our very own backyard. Who knew that a bunch of scrappy, random space debris could eventually form into these diverse, intricate wonders?

For years, scientists had a pretty good idea of what our own solar system's planets were made of based on observations and measurements. But then, in 1995, the world was rocked by the discovery of the first exoplanet orbiting a star outside of our Solar System. And since then, more than 4,000 other planetary wonders have been discovered, each with its own unique story. From their intricate and diverse geologic, chemical, and biological processes to the mysterious forces that shaped them into the worlds they are today.

So, let's explore the mystery of how planets form and even take a trip to some of these exotic exoplanets.

But right now, though, I'm going to talk about the planets very close to our home - in our solar system.

My Very Excellent Mother Just Served Us Noodles

Are you ready to have some fun with the planets? Let's bring back those childhood memories of the classic mnemonic device taught in 5th grade "My Very Excellent Mother Just Served Us Noodles."

Do you remember what it stood for? That's right! It's the order of the planets in our solar system, starting from the closest to the Sun, all the way to the farthest away. And yes, I know there are other mnemonic devices out there, but let's stick to this classic one, shall we?

So, let's break it down. The first letter of each word represents the first letter of the planet's name. For example, the first word, "My", represents Mercury, the planet closest to the Sun. Then comes the word Very representing Venus, second out from the Sun - and there you have it folks! The 8 planets in order, starting from the closest to the Sun:

Mercury, Venus, Earth, Mars, Jupiter, Saturn, Uranus, and Neptune!

"My Very Excellent Mother Just Served Us Noodles."

Don't you feel smarter already? Ha ha, let's keep exploring the wonders of the universe!

Each of these planets exhibits extraordinarily different characteristics, and I will be talking about these characteristics in a while.

My Very Excellent Mother Just Served Us Noodles

Picture our Solar System

So, what does our solar system look like? Let's imagine that we are looking at a model and standing away from it at a distance.

Our solar system is a huge, spinning, flat disk where all the planets are groovin' in almost the same plane, known as the "invariable plane." Imagine four and a half billion years ago, the dust and debris were just chillin' in the cosmos, doing their own thing. But then, they started to get all cozied up, and before you knew it, they were spinning together in the same direction. The result? A pancake-flat disk with the Sun as the disco ball in the center, attracting all the planets to dance around it with its gravity.

Get ready to learn about the cosmic dance-off of the solar system! And let me tell you, each planet has its own unique moves! I can't wait to show you their eccentricities in a bit.

Our telescopes have become increasingly more sensitive over the years. They can show us what is occurring in the far distant reaches of our galaxy and beyond. What I have described to you about how our own solar system was formed applies to all solar systems forming and those in the process of forming.

Imagine a cosmic camera that's zoomed into the furthest reaches of the galaxy and beyond. The picture it captures is mind-blowing!

Now, if you think our solar system is unique, guess what? It's not! The formation process of our solar system is playing out in other parts of the galaxy too! It's like a cosmic factory producing solar systems left, right, and center, just like us. So, remember, next time you look up at the night sky, what you're seeing is just a small part of the big cosmic picture!

Come with me on this thrilling adventure aboard two loveable spacecraft so I can give you an idea of the overall size of our solar system.

Chapter Nineteen

Interstellar Pioneers: Tracing the Epic Journey of the Two Voyager Spacecraft through the Uncharted Territories of the Outer Solar System

Let's talk about the dynamic duo of space exploration - Voyager 1 and Voyager 2! NASA launched these bad boys back in '77 with a pretty sweet mission - to go where no spacecraft has gone before and explore our solar system's outer planets!

Their job? To get up close and personal with Jupiter, Saturn, Uranus, and Neptune and report back all their juicy deets. These spacecraft were equipped to study everything from magnetic fields, atmospheres, and even the rings around the planets. Talk about impressive!

But here's the kicker - these spacecraft weren't just sitting around waiting for their target planets to come to them. Oh no, these guys took advantage of a rare alignment of the outer planets and used their gravity to slingshot their way across space and get to their destinations faster.

Voyager 1 spacecraft, in 1977 awaiting launch. 45 years later it is traveling at 11 miles per second away from our solar system

By 1998, Voyager 1 became the most distant human-made object, 20-plus years on, and in 2012 it entered interstellar space, 11 billion miles from the Sun! The other spacecraft, Voyager 2, continued on to Uranus and Neptune, sending back stunning photos that had everyone gawking.

But get this, they didn't just send it out there empty-handed! Nope, they included something way cooler than your average time capsule - a golden record!

Now, this wasn't just any old record. It was a specially designed golden record that contained sounds and images that were carefully chosen to represent the diversity of life on Earth.

First off, let's talk about the music selection. They didn't just choose the latest Top 40 hits. Oh no, they went for something timeless, something that would make aliens say, "Wow, these Earthlings have taste!" They included a range of music, from Beethoven to Chuck Berry. That's right, Chuck Berry's "Johnny B. Goode" made the cut. Who knows, maybe aliens will start a rock band after hearing that one.

But it's not all about the music. They also included greetings in 55 different languages. And no, it's not just "hello" and "goodbye." They went all out with phrases like "We come in peace" and "May peace prevail on Earth." They even included the sound of a kiss. Yes, you read that right, a kiss. I guess they wanted to make sure aliens know we're a friendly bunch.

The Golden Record also has a bunch of sounds from Earth, like thunder, volcanoes, and animals. Because who doesn't love the sound of a humpback whale? Plus, they included the laughter of a human, which is honestly one of the best sounds in the world.

They included pictures of our planet, as well as people, animals, and even diagrams of our DNA. That's right, aliens, we're complex beings.

All in all, the Golden Record is like a mixtape from Earth to the universe. It's a snapshot of who we are and what we're all about. It's like saying, "Hey, aliens, this is us, and we're pretty cool. Wanna hang out?"

The record also included a map that shows the location of our planet in the galaxy, just in case any extraterrestrial beings stumble upon it. And, get this, the record was made out of gold-plated copper, which is basically the fanciest metal you can get!

So, why did NASA send a golden record into space? Well, it was a way to share a piece of our world with any intelligent life that may be out there. Plus, it's just plain cool to imagine our music and culture floating out there in the vast expanse of the universe.

The Voyager spacecraft is out there, cruising through the cosmos with a golden record on board, just waiting to be discovered by someone (or something) out there. Who knows, maybe one day we'll get a response and start chatting with aliens about our favorite music and movies.

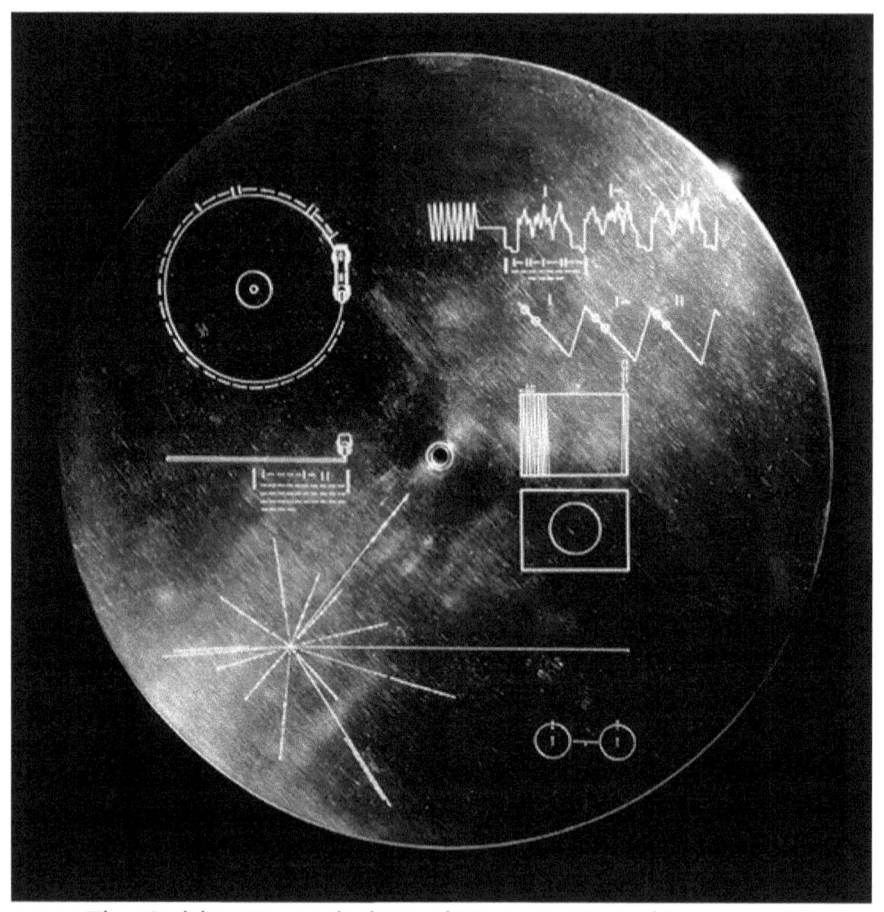

The Golden Record aboard Voyager 1 and Voyager 2

And believe it or not - even to this day, the two spacecraft are still sending data back to us Earthlings. These guys are the Energizer bunnies of space exploration - they just keep going and going.

But the journey doesn't end there! The Voyager spacecraft will continue their adventure for the next billion years - heading towards their final destinations. Voyager 1 is headed towards a star named AC+793888, which it will reach in 40,000 years. And Voyager 2? It's on its way to Ross 248 in the Andromeda Constellation.

So just think, it took these spacecraft 45 years to reach the outer limits of our solar system, traveling at a lightning speed of 11 miles per second. I'll repeat that slowly – they are traveling at a speed of 11 miles each second. Can you even wrap your mind around that? I'll give you a moment to try (take all the time you need, we can wait!)

Slowly, count to 5!

You get the idea!

That was a 5-second pause – the spacecraft have just hurtled 55 miles through space!

They were never designed to travel such distances, they were just meant to take a little stroll through Jupiter and Saturn and then call it a day, but boy, did they deliver! They're still beaming data back to Earth and causing scientists to rethink their theories about the protective effects of the Sun's magnetic field from harmful cosmic rays and dust - something they weren't really designed to do. It's pretty wild to think about - I mean, these two spacecraft are like the tiniest ants in a universe-sized picnic!

Who knew that such a small spacecraft could have such a big impact on our understanding of the Universe?

Now, who's ready to blast off into space to meet our neighbors?

An Intimate Tour of the Diverse and Fascinating Worlds that Orbit Our Sun

Mercury

Hold on to your seat folks, because we're about to blast off to the speedster of our solar system, the planet Mercury! This little guy is hurtling around the Sun at a lightning-fast pace of 29 miles per second. With an orbit around the Sun that takes a mere 88 days compared to Earth's plodding 365 days, Mercury is the closest planet to the Sun and the smallest of them all. It's about the same size as our Moon, with a rocky surface pockmarked by billions of years of planetary bombardment. Mercury rotates on its axis once every 55 of our days, which is like watching paint dry compared to Earth's 24-hour rotation. And speaking of life, you won't find any on this planet - its thin atmosphere and extreme temperatures make it a no-go zone. If you could stand on Mercury, the Sun would appear a staggering three times larger than it does from here on Earth. Despite its proximity to the Sun, Mercury is one of the least explored planets in our solar system.

Mercury is the closet planet to the Sun

NASA has sent two spacecraft to circle the planet, and the European Space Agency has launched three probes that have flown by several times. Scientists are trying to figure out if Mercury has a magnetic field, what's causing the mysterious hollows on its surface, and why there's ice at its poles despite the scorching temperatures. These missions will help us better understand the formation of our solar system and uncover more of Mercury's secrets.

Venus

Let's embark on a journey to Venus, the fiery goddess of the solar system! Nestled between Earth and Mercury, this hot tamale of a planet is about the same size as Earth, but don't let that fool you. Venus is about as hospitable as a lion's den!

The surface of Venus is covered in dome-like volcanoes, rifts and mountains, great volcanic plains, and ridged plateaus. Its toxic clouds whip around the planet at 200 miles an hour, with immense crushing pressure. Probes that have been sent to this beautiful-looking world have not survived more than a couple of hours because it's so darn hot- around 900 degrees Fahrenheit, hot enough to melt lead.

With temperatures hot enough to melt lead and an atmosphere thick with carbon dioxide and sulfuric acid clouds that smell like rotten eggs, Venus is definitely not a place you'd want to call home. But that's not stopping NASA and the European Space Agency from sending three brave probes - Veritas, DaVinci, and EnVision - to explore this hellish world.

These intrepid probes will be gathering data on how Venus lost its potential to become an inhabited world, taking high-resolution images of its surface, and one of them will even make a one-hour descent to collect valuable data before it meets its fiery demise. Venus spins like a slowpoke, taking 243 Earth days for one rotation, and rotates in the opposite direction to the rest of the planets. That means sunrises are in the west and sunsets in the east.

And just when you thought things couldn't get any crazier, some scientists believe that any potential life on Venus could be up in the clouds, thriving as microbes. Will these daring probes be able to solve this mystery? Only time will tell!

Earth - Home Sweet Home

Okay, let's talk about this little blue marble we call Earth. It's not the biggest planet in the solar system, but it's definitely got a big personality. Why? Because it's the only one that has life.

So, let's talk about what it takes for life to party on a planet. First off, you gotta have some real estate - and Earth's got it in spades. At a modest 7,918 miles in diameter, it's the fifth largest planet in our solar system. But size isn't everything. What really sets Earth apart from the other planets is its ability to throw the ultimate rager: hosting all kinds of crazy life forms.

But wait, what exactly do you need to have a party? Well, Earth's got it all. A solid atmosphere, comfy temperatures, and plenty of water to keep the drinks flowing. Seriously, 71% of Earth's surface is covered in water – salty water filling a lot of swimming pools! And then there's the freshwater, although it's not always easy to find. Plus, Earth has a sweet magnetic field to keep the riffraff from the sun's radiation.

But let's get serious for a minute. Life ain't just a party - it's also a struggle. And Earth's seen some real changes over the years. From ice ages to a dinosaur killer asteroid , this planet's been through some wild times. But even with all that chaos, life has managed to roll with the punches and come out on top. It's like the ultimate game of "Survivor" - except instead of voting each other off the island, the contestants adapt and evolve to keep their spot in the game.

And speaking of evolution, Earth's got some real showstoppers in the brain department. We're talking about over 100 billion neurons processing more information than a Hollywood agent during awards season. This is what sets humans apart from other animals - and all thanks to our big ol' brains.

Our brains are like a supercomputer on steroids, capable of processing more information than you can shake a stick at. It's responsible for all the cool stuff we do, like thinking, feeling, learning, and creating. It's the ultimate power tool, and it's all ours.

So, to sum it up - Earth's got all the ingredients for the ultimate party: size, water, atmosphere, magnetic field, and of course, life. It's the ultimate hostess with the mostess. And while other planets might try to compete, Earth's the OG party planet - and we're lucky enough to call it home.

Sure, other planets might have a few of these things, but none of them have the killer combo that makes Earth the ultimate spot in the solar system.

Is there a chance that there's some other form of intelligent life lurking out there in the vast expanse of the universe? Maybe some aliens who are even more advanced than us? Or are we just the coolest kids on the cosmic block? The answer to that question is... drumroll, please... we don't know! Yup, it's a total mystery. But hey, that just leaves the door wide open for all sorts of wild speculations. Maybe there are aliens with tentacles for hair, and noses on their toes! Or maybe they communicate telepathically, and their favorite food is peanut butter and jelly sandwiches. Who knows? One thing's for sure, though - it's a big, weird universe out there, and there's always a chance that we might stumble across something truly amazing.

Man on the Moon

Are you ready to blast off on a journey to the Moon?

First things first - let's talk about our Moon's appearance. It's basically a giant cheese wheel floating in space, right? Wrong! Contrary to popular belief, the Moon is not made of cheese (sorry, Wallace and Gromit). It's actually made of rock and dust, which is way less delicious, but way more scientifically accurate.

But where did the Moon come from you ask? Well, there are a few theories, but the most popular one involves a giant collision between Earth and a Mars-sized object. That's right - the Moon was basically created by a cosmic car crash. Talk about a fender bender of epic proportions!

So, what's the Moon like up close and personal? Well, it's pretty dusty (thanks a lot, Apollo astronauts), and it's also super cold. Temperatures on the Moon can range from a balmy 260 degrees Fahrenheit during the day to a mighty chilly minus 280 degrees Fahrenheit at night.

Speaking of the Apollo missions, did you know that twelve humans have walked on the Moon? That's right, a dozen brave souls have taken a giant leap for mankind and left their footprints on the lunar surface. I like to think that they all did a little bit of moonwalking while they were up there. You know, for the 'gram.

But the Moon isn't just a cool place to visit - it also plays an important role in our ecosystem. Its gravitational pull causes the tides in our oceans, which has a huge impact on marine life and coastal ecosystems. Plus, its presence stabilizes Earth's axial tilt, which helps to create a stable climate for life to thrive. So, basically, the Moon is like the responsible friend who always makes sure you get home safe at the end of the night.

And the Moon isn't done yet! NASA and other space agencies are currently planning a return to the Moon as part of the Artemis program. The goal is to establish a sustainable presence on the lunar surface and pave the way for future human exploration of Mars and beyond. Who knows, maybe someday we'll all be living in moon bases and commuting to work via rocket ship. I, for one, can't wait for the lunar commute - it beats rush hour traffic any day of the week.

The Moon has been hanging out with us for billions of years, and it's not going anywhere anytime soon (unless something really catastrophic happens, but let's not dwell on that). Next time you look at that big, beautiful cheese wheel...er, I mean, rock, give it a little wink and say, "Hey, Moon, you rock!" And who knows - maybe it'll wink back.

Mars

Let's hop over to the cool kid of the solar system, Mars! This fiery planet is orbiting the Sun on the outer side of our own orbit, making it our next-door neighbor in space. And let me tell you; scientists have been all up in its business for the past 45 years with a bunch of fancy-pants rovers and orbiting spacecraft. Mars might as well be on Tinder with how many times it's been swiped right. https://mars.nasa.gov/mars2020/participate/sounds/?playlist=mars&item=mars-helicopter-flying&type=mars

Now, let's address the elephant in the room. We've all had that one friend who's convinced little green men are living on Mars. But sorry to burst your bubble, folks. Science has thoroughly explored this planet and found zero evidence of any past or present life forms. Sure, there might have been some habitable conditions back in the day when liquid water was on the surface, but that doesn't mean there's anyone (or anything) home.

But don't let that dampen your excitement for this red-hot planet. When you're stargazing and see a bright, reddish twinkle, you're looking at Mars, baby! That red hue is all thanks to a thick layer of oxidized iron dust and rocks on its surface. Think of it as rust on steroids. It's like Mars skipped leg day and just built up a massive layer of rust instead.

Mars has some impressive geological features. This planet doesn't have tectonic plates like Earth, which means no earthquakes. But it makes up for it with massive volcanoes that would make even Mount Vesuvius quake in its boots. The Olympus Mons volcano on Mars is so big, it makes Mount Everest look like a molehill. It is 13 and a half miles high - Everest is 5 and a half miles high. And how about the Mars' Grand Canyon - it's five times longer and more than twice as deep as the one we have here on Earth. Talk about a grandiose canyon.

Mars has had visits from 130 robot probes since 1997

Now, Mars does have an atmosphere but it's thin as a supermodel, with about 100 times less air pressure than Earth's atmosphere. But don't let that fool you - Mars still knows how to party. It's got some wild weather events and serious dust storms that can envelop the entire planet. And if you're looking for some snow, you won't find that stuff here. Nope, it's all about the carbon dioxide ice on Mars.

Oh, and did I mention how chilly it can get? On average, we're talking minus 80 degrees Fahrenheit, with the poles dipping down to a bone-chilling minus 195 degrees Fahrenheit. But hey, it's not all bad - near the equator, you can bask in a toasty 70 degrees Fahrenheit.

Think of it as Mars being that friend who can't decide whether to hit the beach or the ski slopes. And the orbit of Mars is all over the place, taking it farther away from the Sun and then bringing it closer again, called an eccentric orbit. This makes for some pretty fluctuating seasons longer than we experience on Earth.

We've got 13 probes snooping around the Red Planet, but let's be real here, it's not all rainbows and sunshine. Since 1977, there have been more rocket flops than touchdowns, with a whopping 28 failures and 19 successes. Ouch! But hey, nobody said rocket science was easy, right?

Fast forward to today, and NASA's been killin' it with their high-tech gizmos, including the Phoenix Lander, the MAVEN orbiter, the long-lived Opportunity and Spirit rovers, and the big daddy of them all, Curiosity. This a beast of a robot, tips the scales at over one ton and boasts some serious scientific swag. Landing on the surface back in 2011, it was only supposed to last a couple of years, but ten years later, it's still going strong! It's like the Energizer bunny, but on steroids! And those pictures it sends back to Earth are pure gold.

Enter the Perseverance rover, launched in 2021, with a mission to sniff out signs of ancient life and scoop up some rocks and soil to bring back to Earth. The landing spot was a doozy, a former river delta turned lake bed, with scientists thinking it's prime real estate for finding some old microbial buddies. And get this; they're recording sound on this mission! Yeah, you heard me right; they're blasting rocks with lasers and recording the sounds it makes to determine what kind of rock it is. That's some next-level stuff right there.

Mars has two little moons, Phobos and Deimos. Asaph Hall, a skeptical astronomer, discovered them back in 1877 with the help of his trusty telescope and his wife's encouragement. And one of those craters on Phobos? Yeah, it's named after his wifey, Stickney. Cute, huh?

But let's face it; Mars isn't always a picnic. Dust storms can sweep in and ruin a good day's work, and communication with the equipment can be spotty at times.

Listen to this sound recording of the laser beam as it zaps the rock...
https://mars.nasa.gov/mars2020/participate/sounds/?playlist=mars&item=laser-zaps-rock-01&type=mars

Here is another recording of the rover moving over the rocks...
https://mars.nasa.gov/mars2020/participate/sounds/?playlist=mars&item=rover-driving-filtered&type=mars

This is the sound of the Ingenuity helicopter flying over the surface of Mars..
https://mars.nasa.gov/mars2020/participate/sounds/?playlist=mars&item=mars-helicopter-flying&type=mars

And this is the sound of the wind...
https://mars.nasa.gov/mars2020/participate/sounds/?playlist=mars&item=wind-on-mars-01&type=mars

Jupiter

Hold onto your space helmets and buckle up because we're headed to the fifth planet from the Sun. Jupiter is its name, and it's one of two gas giants orbiting the Sun, the other being Saturn. But let's be real here, Jupiter is the big cheese, the top dog, and the boss planet of the solar system. Its size is so ginormous, you could combine the masses of all the other planets, and it still wouldn't match up to Jupiter's twice-the-mass-of-that status.

You want a size comparison? Picture a grape representing Earth and a basketball that's Jupiter. Yeah, that's right, it's that big! And get this, Jupiter's heart is probably a solid Earth-size core surrounded by hydrogen and helium gases. This planet is a heavyweight champ. If you weigh 200 pounds on Earth, you'd weigh a whopping 480 pounds on Jupiter, all thanks to its crushing gravity.

Jupiter's Red Spot is a storm that has been raging for hundreds of years and is about the size of Earth

Imagine a giant, gas-filled beach ball painted with the wildest colors you can imagine. Jupiter's got stripes, baby - and we're not talking about the kind you wear on your shirt. These stripes are like Jupiter's own personal racing lanes, and they're filled with all sorts of crazy weather patterns. It's like the planet's been on a bender and is feeling the effects of a serious hangover.

But that's not all - Jupiter's got spots, too. And no, we're not talking about acne. These spots are like freckles on a sun-kissed beach babe. They're just part of Jupiter's charming personality.

Jupiter is famous for its Great Red Spot, which is basically a giant continuous raging storm that has been going on for hundreds of years. It's so massive, it's the size of Earth, but it does fluctuate in size. And all around it are swirling, windy clouds of cold ammonia crystals and water floating in an atmosphere of hydrogen and helium. It's a hostile environment, folks, and not exactly a place where life could thrive.

But that doesn't stop us from exploring it! Nine spacecraft have visited Jupiter, including the first, Galileo, in 1995, which dropped a probe into the atmosphere and found it to be surprisingly dry, despite the abundance of water near its equator. The current spacecraft orbiting Jupiter is named Juno, and since its arrival in 2016, it's been providing us with hundreds of remarkable close-up pictures of the surface. One of the coolest pictures is of the north pole, where you can see a beautiful and seemingly artistic arrangement of 8 Earth-size storms, or cyclones, that spin in the same direction, encircling a central one. And get this, those cyclones are also present at the south pole, spinning continuously and never merging. Talk about weather patterns!

And if that's not enough to blow your mind, Jupiter is a speedy planet. Compared with Earth's orbit of 24 hours, Jupiter spins about two and a half times faster. And while Earth takes one year to circle the Sun, Jupiter takes its sweet time, circling in 12 Earth years. Plus, there's a visible equatorial bulge in Jupiter's shape, and you can see that the planet is squashed even through an amateur telescope. Because it's spinning so fast, it bulges around the middle and is actually the fastest-spinning body in the solar system. Who knew a gas giant could be such a speed demon?

So, Galileo saw four moons when he trained his telescope on Jupiter in 1610. Since then, 79 moons have been seen orbiting the planet, and more will likely be discovered. And get this; faint rings accompany the moons on their journey around the Sun. How cool is that?

But as much as we love Jupiter, we must admit; it's a bit of a bully. Because of its massive size, it has an immense gravitational influence on objects that come too close to it. In fact, it has some effect on the whole solar system. In the distant past, many celestial objects flying too close to Jupiter were mercilessly drawn in by its gravitational pull and were destroyed. Imagine being a poor little asteroid, just minding your own business

The Moons of Jupiter

Let's talk about the biggest moons of Jupiter that Galileo could spot with his humble telescope back in the 1600s. But now, with a good pair of binoculars, you can spot them too! These four moons, Io, Europa, Ganymede, and Callisto, rotate around the planet on the same plane as its equator.

Ganymede

Ganymede is one of Jupiter's moons that's larger than Mercury and as light as a birthday balloon. That's right - Ganymede is the big kid on the block - the largest moon in the solar system. But don't let its size fool you; this moon is a real lightweight when it comes to atmosphere. It's mostly made up of oxygen and ozone, so don't go holding your breath for fresh air up there. So, if you want to hang with the cool kids, forget about those other moons and get ready for Ganymede to take center stage. It's big, it's bold, and it's got more secrets than Area 51. Ganymede, baby!

But what really makes Ganymede stand out from the rest is its pimples, grooves, and ridges. It's like its surface got hit with an intergalactic pizza party, and we're not talking about just any ordinary pizza, no sir. We're talking about the kind of pizza that comes from a cosmic oven baked with tidal forces generated by Jupiter's gravity. These forces keep Ganymede's tectonic plates moving and shaking, making it the most active and dynamic moon in town.

But Ganymede has a secret, and it's juicier than a gossip mag. Hidden beneath its icy surface, there's a salty ocean that could make a water park jealous. Rumor has it that it has more water than Earth, and some scientists even think it could be hiding some extraterrestrial life forms. That's why the European Space Agency is planning a mission to Ganymede in 2030, and they're not just going there for the view.

Io

Get ready to meet the moon that's hotter than a jalapeño on fire - Io! This little volcanic rock orbits Jupiter and is like a hot tamale with over 400 active volcanoes - that's more than any other moon in the solar system! These volcanoes are so intense they make volcano on Earth look like tiny hills, and their sulfur plumes shoot up hundreds of miles into the air. Talk about a bad hair day! But Io's not just a fiery diva. It's also got some serious curves, thanks to the tidal forces created by its dance with Jupiter and its moon pals, Europa and Ganymede. These forces heat up Io's insides to a scorching 3,000 degrees Fahrenheit - hotter than a summer day in Phoenix, Arizona. But don't worry; there's also a cool side to Io with icy fields of carbon dioxide that could freeze your nose hairs off at minus 200 degrees Fahrenheit. It's like Io is trying to be a cosmic yin and yang all at once. And if you think Io is slow and steady, think again. This moon is zooming around Jupiter like a speed demon, taking less than two Earth days for a full orbit. It's like it's running a marathon with Jupiter cheering it on like a proud coach. Plus, just like a model, Io always shows off its best side to Jupiter - talk about being a celestial diva! So next time you're feeling hot and cold at the same time, think of Io and its fiery volcanoes and icy fields. Who knew a moon could be such a spicy and cool combination?

Europa

Europa is the ocean moon of Jupiter. It's the prime spot where astrobiologists think they could uncover some alien life forms. And where would these little rascals be hiding? In the salty waters that span for miles and miles.

Don't let Europa's frozen surface fool you because beneath it lies a water wonderland deep enough to make the Mariana Trench look like a kiddie pool. It is frozen about 10-15 miles deep. Scientists believe there's a salty ocean below the ice just waiting to be explored, about 40 to 100 miles deep, and that's not even counting the 100-mile-high plumes of water that have been spotted shooting out from the surface. Future space probes would be able to capture some of this water by swooping close, and the water could then be analyzed.

But hold onto your data pads, folks, because we're not done yet. The cherry on top? Finding some form of life on Europa would be like discovering a needle in a haystack; only this needle would prove that life can actually develop in places other than our little blue planet. That's like finding out your neighbor's dog can talk; it's a game changer, people!

Callisto

Callisto is the fourth largest moon orbiting the gas giant Jupiter. Now, let me tell you, this ain't your run-of-the-mill, plain Jane kind of moon. No, sir! Callisto is a mysterious celestial body shrouded in an aura of intrigue and wonder. With a diameter of nearly 3,000 miles, it's no shrinking violet.

In fact, it's the second-largest moon in the solar system, and it's covered in a thick crust of ice and rock.

But that's not even the most interesting part! Callisto's surface is pockmarked with thousands of craters, some of which are over 100 miles wide. It's like a cosmic pin cushion! And if you think that's impressive, get this: scientists believe that beneath all that icy armor, there may be a subsurface ocean of liquid water that could potentially harbor extraterrestrial life! That's right, folks, we're talking about aliens!

Now, I know what you're thinking. "How can a moon that's so far away from the Sun, and has a surface temperature that's colder than your ex's heart, have liquid water?" Well, let me tell you, it's all thanks to a little something called tidal heating. You see, Callisto's orbit is influenced by not one, not two, but three of Jupiter's other moons, including the fiery and volatile Io. And all that gravitational tugging and pulling generates a whole lot of heat, which could potentially melt the ice and create a habitable environment for all sorts of funky and fascinating creatures.

So there you have it, folks. Callisto may be a distant and cold moon, but it's got a whole lot of personality and mystery to spare. Who knows what other secrets it's hiding just beneath the surface? One thing's for sure, though. We're going to need a bigger rocket if we want to explore all of its wonders.

Jupiter has a whopping 79 moons, and we've only given names to 53 of them. That means there are still 26 moon-sized mysteries waiting to be explored. Some are small and wonky, while others are like giant space rocks that Jupiter collected like seashells on a beach. And who knows what other secrets these moons are hiding? Maybe there's a moon made entirely of cheese or a moon with a secret underground casino where aliens go to gamble away their space bucks. The possibilities are endless, my friends!

Uranus

Now, come with me, and let's fly onward and outward from Earth to another gas giant – Uranus.

Picture a humongous disco ball, but instead of being covered in mirrors, it's covered in swirling blue and green clouds. It's like the planet is constantly busting out some sick dance moves, even though it's floating in the middle of space.

And those rings? They're like the planet's bling, made up of all the rocks and debris that the Uranusians have collected over the years. The rings are so thin; they're practically invisible - kinda like that cheap bracelet you got from the vending machine as a kid.

As for the planet's color, well, let's just say that Uranus is feeling pretty blue these days. It's like the planet's been listening to Adele on repeat and crying its eyes out. But don't worry Uranus - we still think you're cool!

And those hazy bands of clouds? They're like the planet's own personal vape clouds. It's like Uranus is trying to look edgy and mysterious, but it's not fooling anyone. We see you, Uranus, and we appreciate your efforts to stay on-trend.

Those vape clouds are primarily composed of ices such as water, methane, ammonia, rock, and metal. It's the third-largest planet in our solar system, after Jupiter and Saturn, but it's a bit of an oddball in several ways.

For starters, Uranus is tilted at an angle of about 98 degrees, which means it rotates on its side. Imagine trying to walk around with your head tilted at a 98-degree angle - it's not the most practical or comfortable position to be in! Scientists aren't entirely sure how Uranus got tilted like this, but it's thought to have been caused by a collision with a planet-sized object early in its formation.

Another interesting fact about Uranus is that it has the coldest temperature of any planet in our solar system, with an average temperature of minus 195 degrees Celsius (minus 320 degrees Fahrenheit). Brrr!

But get a load of this - Uranus also has a unique feature called the "ice giant weather," which refers to the planet's extreme winds and atmospheric conditions. The winds on Uranus can reach up to 560 miles per hour, making them the fastest winds of any planet in our solar system. It's basically the planet equivalent of a hurricane on steroids!

Discovered all the way back in 1781 by Sir William Herschel, Uranus has only been visited by one spacecraft, NASA's Voyager 2, which flew by the planet in 1986. There are plans in the works for future missions to Uranus, including a proposed orbiter and even a possible mission to explore its moons.

Uranus may be a bit of a cosmic wallflower, but it still knows how to work its unique look. Keep dancing, Uranus, and don't let anyone dull your sparkle.

Saturn

This giant gas planet is like Jupiter's cousin, made up of hydrogen and helium, but it's got something that Jupiter doesn't have: the most amazing, beautiful rings you've ever seen! Seriously, if Saturn were a jewelry store, it would be the fanciest and most unique one out there.

Now, these rings aren't just made up of some boring old material. Oh no, they're composed of billions of pieces of comets, asteroids, and moons that smashed into each other and got captured by Saturn's gravity. Some of these pieces are so small they're like grains of sand, while others are as big as mountains, all circling around Saturn at different speeds.

The ring disc stretches out about 175,000 miles away from the planet, but it's only 30 feet thick on average. That's like a pancake that's been stretched out over an entire football field!

Scientists are still pondering how these rings formed. One theory is that a moon that was orbiting Saturn got smacked by another celestial body and shattered into bits. One of Saturn's moons, Enceladus, has even been seen spewing out ice from a volcano above its hidden ocean at the southern pole, adding to the ring formation. Imagine a hundred volcanoes going off at once - that's what's happening on Enceladus! And not just there - other moons like Titan have cryo-volcanoes erupting with frozen water, ammonia, and methane, all contributing to the ring material.

Now, if ancient humans had telescopes and were able to see Saturn, they probably would have just seen a plain old planet without any rings. But lucky for us, we get to witness this spectacular sight! However, the rings won't last forever. Eventually, all the material that's adding to the rings will run out, and the rings will dissipate and be dragged down to Saturn's surface. So, cherish these rings while you can, folks!

Speaking of the planet itself, Saturn is massive - you could fit 760 Earths inside it! Despite its size, though, its density is such that it would actually float in a big enough bathtub. And get this: if you weighed 100 pounds on Earth, you'd only weigh 10 pounds on Saturn. That's like a dream come true for anyone trying to lose weight!

And Saturn is no couch potato either. It spins around really fast, taking just ten and a half hours to complete one rotation. That causes it to bulge out at the equator like it's got a little bit of a belly. And those winds blowing around the planet? They're whipping around at a whopping 1100 miles an hour! That's like a race car driving around the equator.

Saturn has its fair share of moons, 82 in all, and still counting

Saturn is also a bit of a diva. It orbits around the sun every 29 years, and it's the farthest planet from the sun. But that doesn't stop it from shining brighter than any other planet in our sky, with its stunning rings stealing the show.

With 53 moons already spotted and another 29 waiting to be verified, Saturn is like a moon magnet! It's got a total of 82 moons, and they're all circling around the planet like partygoers at a rave, along with the particles of those seven rings arranged in sections with several gaps. It's another cosmic dance party up there!

The Cassini spacecraft deserves a standing ovation for all it accomplished during its 20-year mission to Saturn. It orbited the planet hundreds of times over the course of 13 years and even flew through the rings in all directions, transmitting data constantly. Thanks to Cassini, we now know a lot more about Saturn and its moons. For example, Titan is basically like Earth, with lakes, rivers, channels, dunes, rain, clouds, mountains, and volcanoes. In fact, the Huygens probe released by Cassini even landed on Titan in 2005, making it the first time we landed on an object in the outer solar system! And get this - scientists think there could even be weird life forms on Titan that rely on liquid hydrocarbons instead of water. That's like a whole new kind of alien!

The Larger Moons of Saturn

Some of the largest and most well-known moons of Saturn include: Titan, Enceladus, Mimas, Tethys, Dione, Rhea, Lapetus, Hyperion

First up is **Titan**, the largest of Saturn's moons. Titan is so big that it's actually bigger than the planet Mercury! It's also the only moon in the solar system with a thick atmosphere. This means that if you were to stand on Titan's surface, you'd need a spacesuit to survive. But don't worry, Titan's atmosphere is mostly nitrogen, so at least you'd have plenty of air to breathe.

Next up is **Enceladus**, known for its icy surface and towering geysers. The geysers shoot out icy particles, which form Saturn's famous E ring. It's almost like a giant cosmic soda fountain! In fact, Enceladus is one of the most active places in the entire solar system. So, if you're ever thirsty in space, just head on over to Enceladus and grab yourself a refreshing drink!

Moving on, we have **Mimas**, which is often called the "Death Star Moon" because it resembles the famous Star Wars weapon. Now, I don't know about you, but I wouldn't want to mess with a moon that looks like it could blow up an entire planet. But fear not, Mimas is actually quite small, so the chances of it destroying anything are pretty slim.

Then there's **Tethys**, which is a moon that's made mostly of ice. It's also home to a gigantic impact crater called Odysseus. If you were to stand at the center of the crater, you'd be surrounded by walls that are over 2 miles high! That's taller than any building on Earth. So, if you're ever feeling small and insignificant, remember that there's a crater on Tethys that's bigger than most mountains.

Next, we have **Dione**, which is another icy moon. It's not as active as Enceladus, but it does have some interesting features, like a long, winding valley called Padua Chasmata. It's like the Grand Canyon, but colder and with fewer tourists.

Moving on to **Rhea**, which is a moon that's covered in craters. Some of them are so big that they have their own little moons! Imagine being a moon that has its own moon. Talk about being the ultimate celestial boss.

Then we have **Lapetus**, which is a moon that's half dark and half light. It's almost like someone took a giant space paintbrush and painted half of it black. Lapetus also has a giant ridge that runs around its equator. It's like the moon has a belt, but instead of holding up its space pants, it's just there for looks.

Lastly, we have **Hyperion**, which is a moon that looks like a giant sponge. It's covered in weird, crater-like holes that make it look like a piece of space cheese. It's also one of the largest irregularly shaped objects in the solar system. So, if you're ever feeling a little lumpy or misshapen, just remember that there's a moon out there, that looks like it got hit with a giant space cheese grater.

Saturn's largest moons are a diverse and fascinating bunch. Each one has a unique personality, from the Death Star Moon to the Space Cheese Moon. So, the next time you're stargazing, take a moment to appreciate these amazing celestial bodies. And who knows, maybe one day we'll get to visit them in person and see their quirks up close and personal!

Neptune

Neptune, the distant blue giant of our solar system, is a fascinating and mysterious planet that has captured the imagination of scientists and stargazers alike.

First things first, let's talk about Neptune's size. It's a big boy, coming in as the fourth largest planet in our solar system, and it's so huge that you could fit 57 Earths inside it! That's a lot of Earths. It's also very far away from the Sun, which means it's cold enough to make even the most seasoned polar explorer shiver in their space boots.

But don't let the frigid temperatures fool you; Neptune is a hotbed of activity. It's got some of the wildest weather in the solar system, with storms that can last for decades and winds that can reach speeds of up to 1,300 miles per hour! That's faster than a cheetah on a rocket-powered skateboard.

One of the most prominent features of Neptune is its stunning blue color, which is caused by the presence of methane gas in its atmosphere. Methane absorbs red light, giving the planet a blue hue that's so beautiful that it would make even the most stoic of aliens shed a tear.

Neptune also has a set of faint and mysterious rings orbiting around it. These rings are made up of billions of tiny particles, some as small as a grain of sand, and they're so faint that they were only discovered in 1989 by the Voyager 2 spacecraft. It's like Neptune was hiding a secret, and it took us decades to uncover it.

But Neptune isn't just a pretty face; it's also got some seriously cool moons. There are a total of 14 moons that we know of, and they all have names that sound like they were plucked from a sci-fi movie. There's Triton, Naiad, Thalassa, Despina, and Nereid, to name a few. These moons come in all shapes and sizes, some rocky and others are icy, and they all orbit around Neptune like a pack of loyal space puppies.

Speaking of Triton, it's one of the most coolest, and when I say the coolest moons in our solar system, I really do mean the coolest. Not only is it the largest moon of Neptune, but it's also one of the coldest places in the solar system, with surface temperatures that can reach as low as minus 391 degrees Fahrenheit. That's colder than a polar bear's toenails! Despite its frigid temperatures, Triton is geologically active, with cryovolcanoes that spew out a mix of ice, nitrogen, and dust. It's like a cosmic snow cone machine, but instead of syrup, it's spitting out frozen gases.

The Moons of Neptune

Well, well, well, it seems like Neptune has a whole bunch of watery friends, all with fancy Greek names. Talk about being posh! I mean, what's the point of having a moon if you can't give it a name that'll make it feel like it's living its best life? In conclusion, Neptune may be a distant and frigid planet, but it's also a wonderland of beautiful blue hues, wild weather, and cool moons. So, the next time you look up at the night sky, take a moment to appreciate this giant gas giant and all the strange and beautiful things it has to offer.

Let's start with the closest to Neptune, the little **Naiad**. It's so close that it's basically Neptune's shadow. I bet Naiad is feeling pretty neglected. Maybe they should just rename it 'Shadow' and be done with it.

Thalassa, Despina, and **Galatea** are all named after different water nymphs, but let's be real; they're all just different types of mermaids. Galatea is probably the one with a shell bra and a cool seaweed hairstyle, while Thalassa is more of a traditional, elegant mermaid. Despina, on the other hand, is the one who's always getting into trouble and needs rescuing.

Larissa sounds like a posh lady who lunches at fancy restaurants and drinks tea with her pinkie finger out. But don't let that fool you; she's still a moon of Neptune. S/2004 N1 is just too lazy to come up with a name; I mean, it's only been like 17 years since it was discovered. Get with the program, S/2004 N1!

Proteus sounds like a jock who's always flexing his muscles, but in reality, it's just a moon that's oddly shaped. I guess it's like a jock who skipped leg day.

Triton is the big guy in the group, the one who everyone's a little scared of but secretly wants to be friends with. It's also the only moon of Neptune that's big enough to be rounded by its own gravity, so it's definitely got some bragging rights.

Halimede, Sao, Laomedeia, Psamathe, and **Neso** are all distant moons that no one really pays attention to. I guess it's like they're the wallflowers at a party. But hey, they've got cool Greek names, so that's something, right?

All in all, Neptune's moons are a pretty interesting bunch, with fancy Greek names and unique personalities. But let's be real; none of them are as cool as Triton. Sorry, guys, it's just the truth.

Nereid (nee·ree·uhd)sounds like a total drama queen, but I guess that's appropriate for a moon that's been known to have sudden brightenings and disappearances. Maybe she just needs a little attention.

Nereid is the most fascinating moon of Saturn! She is a real enigma, this one. She is a sea nymph held in Greek mythology to be the daughter of the sea god Nereus. She is the third largest moon of Saturn, with a diameter of 340 miles. Not too shabby, right? But don't be fooled by her size - this little lady has some seriously peculiar habits.

Most of Saturn's moons have a relatively circular orbit, but Nereid is the black sheep of the family with her highly eccentric orbit. It's like an oval shape, meaning its distance from Saturn changes drastically during its orbit. Talk about mood swings! Nereid swings out to almost seven times as far from Neptune at one end of its orbit as the other end. It's like the moon is trying to break a record for the most outrageous orbit ever.

It's also the farthest out of all the major moons of Saturn. This moon takes almost a year and 360 Earth days to complete a single orbit. That's like having a birthday once a year, but for Nereid, it's just another day in the office.

Now, let's talk about Nereid's style. She is so dark; it's almost like she is wearing black to mourn her status as the odd one out in the family. Scientists find her tough to study because of her dark surface, but they believe she is made up of a mix of ice and rock, like many other outer solar system icy moons.

But here's where it gets really funky. Nereid has an irregular shape. She is not your typical round-shaped moon like the others. Instead, she looks like a lumpy, potato-shaped object. Scientists think it's because countless asteroids and comets have pelted Nereid over her long history. So, if you see Nereid walking around with some bruises and bumps, don't be surprised.

Despite all her quirks, Nereid is a fascinating and weird cousin of the Saturnian moon family. She keeps us on our toes, and we're always discovering new things about her. So, let's give a round of applause to Nereid, the black sheep of Saturn's moon family!

In conclusion, Neptune may be a distant and frigid planet, but it's also a wonderland of beautiful blue hues, wild weather, and cool moons. So, take a moment to appreciate this giant gas giant, when you gaze into the dark sky, and all the strange and beautiful things it has to offer.

Pluto

There's this tiny, cute little guy in our solar system called Pluto. He's a bit of an oddball, like the kid who brings a weird lunch to school every day and everyone stares at him in confusion.

Pluto used to be the coolest kid on the block, back when he was considered the ninth planet. But in 2006, the International Astronomical Union (IAU) knocked him down a peg and called him a "dwarf planet." Ouch, that's gotta hurt.

Despite his diminutive size - he's only about two-thirds the size of our Moon - Pluto still manages to pack a punch. He's got an eccentric orbit that takes him from 30 to 49 astronomical units away from the Sun – an astronomical unit is the distance from the Earth to the Sun, that's about 93 million miles. At the closest distance of his orbit, he is two and three-quarter billion miles away from the sun, and at the farthest point of his orbit, he is four and a half billion miles away from the sun. It's an elliptical orbit but with staggering distances, right?

So, when Pluto orbits the sun it's like someone doing a marathon and then stopping for a cup of coffee halfway through.

He's also got a thin atmosphere made up of nitrogen, methane, and carbon monoxide, which is like the planetary equivalent of a fancy cocktail.

Pluto's got some cool features, too. He's got five moons, which is like having your own personal entourage. And his surface is covered in icy mountains, plains, and valleys, like a winter wonderland that never ends. Plus, he's got a reddish-brown color that's thought to be caused by some funky organic molecules called tholins. It's like he's been hitting the spray tan a little too hard.

Some people still argue about whether Pluto should be considered a planet or not, but let's face it - he's like that weird cousin you can't help but love. So, here's to Pluto, the scrappy underdog of our solar system who still manages to hold his own among the big boys.

Chapter Twenty

Space Objects

Asteroids

Well, well, you want to know about asteroids, huh? Let me tell you, these bad boys are out of this world! Literally, so, where do they come from, you ask? Great question. They're basically just floating rocks in space, remnants from the formation of our solar system. Think of them like the crumbs left over after baking a cosmic cake.

Now, what are they made of? Well, it's a mixed bag. Some asteroids are made up of rock, others of metal, and some of both. But don't let their composition fool you; these little buggers can range in size from tiny pebbles to behemoths over 1,000 kilometers wide. That's like having a mountain floating in space!

So, what do they look like? Some look like little potatoes, others like cosmic pancakes, and some even have their own mini-moons orbiting them. Talk about fancy! Well, it's hard to generalize, as asteroids can come in all shapes and sizes.

Now, are they a danger to Earth? Well, let's just say they've been known to cause a bit of trouble from time to time. The bigger ones can pack quite a punch if they collide with our planet, and even the smaller ones can cause some serious damage. But don't worry; NASA's got a handle on this. They're constantly monitoring asteroids and developing plans to deflect them if they get too close for comfort.

So, how fast do they move? Well, it depends on the asteroid's orbit and distance from the sun. Some can travel as slow as a turtle, while others move at speeds faster than Usain Bolt. But regardless of their speed, these rocky objects are definitely a sight to behold.

Asteroids are like the sprinkles on top of our cosmic cake. They come in all shapes and sizes, are made of different materials, and can move at various speeds. While they can pose a threat to Earth, we've got smart people keeping an eye on them, so no need to worry. Just sit back, relax, and enjoy the show!

Superbolides

Now, if you think asteroids are bad news, wait till you hear about superbolides. These suckers are like super powerful firecrackers. They're caused by massive space objects, like asteroids or comets, entering Earth's atmosphere at super-fast speeds. The air in front of them gets so compressed that it heats up and creates a visible trail of light in the sky, also known as a meteor or shooting star. But we get a superbolide when the space object is really large and super-fast, like the Chelyabinsk meteor that hit Russia in 2013. This explosive display in the sky is so bright that it can even be seen during daylight hours! And if it explodes at a low altitude, well, let's just say the results can be, uh, shattering.

But let's get a perspective on these objects. The Chelyabinsk meteor may have been the largest object to enter Earth's atmosphere since 1908, but it was a mere 60 feet in diameter and 10,000 tons in weight. Compare that to the asteroid that wiped out the dinosaurs! That bad boy was estimated to be six to nine miles in diameter and caused massive wildfires, earthquakes, tsunamis, and a "nuclear winter" effect that blocked out the sun's light and caused a significant drop in temperatures worldwide. If that doesn't put things into perspective, I don't know what will.

So, what are we doing to protect ourselves from these cosmic bullies? Well, for starters, we're using supercomputer simulations to estimate the damage that asteroid fragments could cause. We're also sending spacecraft to visit asteroids and learn about them as much as possible. NASA's Galileo spacecraft was the first to visit an asteroid way back in 1991, and since then, several other spacecraft have orbited, landed, and collected samples from asteroids like Eros, Itokawa, Vesta, and Ceres. So, you see, we're not just sitting around twiddling our thumbs, waiting for the next asteroid to hit us. We're taking action, and we're doing it with a sense of humor. After all, laughter is the best medicine, right? Unless you're trying to dodge an asteroid, a good bunker may be the better choice.

Spotting Near-Earth Asteroids

Scientists are on the hunt for invisible asteroids! That's right, these sneaky space rocks are hiding out in the cosmos, and it's up to our trusty astronomers to track them down.

Using super-powered telescopes like the Pan-STARRS1 and the Catalina Sky Survey, scientists can detect even the faintest glimmer of light bouncing off these elusive asteroids. Talk about high-tech detective work!

NASA's NEOWISE spacecraft takes things to the next level by using infrared light to detect the heat emitted by asteroids.

But what about those pesky asteroids that are too dark or too small to be seen by telescopes? No problem - enter radar! By bouncing radio waves off the asteroid, scientists can create a detailed image of its shape and surface features. It's like a cosmic game of Marco Polo!

And get this - even regular folks like us can get in on the action! NASA's Asteroid Hunters program enlists citizen scientists to analyze astronomical data and help identify new asteroids. That's right; you could be the one to discover the next big space rock!

Of course, once an asteroid is spotted, it's not all fun and games. Scientists have to track its orbit to determine if it poses a threat to our beloved planet Earth. So, if you see a bunch of scientists huddled around a computer screen, don't worry - they're just trying to save the world!

In fact, just today, I was reading a headline from the "Scientific American" magazine about the latest asteroid discoveries.

"New Space found Asteroid May Strike Earth in 2046, NASA Says: Preliminary estimates suggest that a 50-meter space rock called 2023 DW has a roughly one-in-600 chance of colliding with our planet in 23 years"

Listen up, folks - we've got an asteroid coming in hot! And by hot, we mean really, really close to Earth. Brace yourselves for February 14, 2046 - Valentine's Day just got a whole lot more exciting. But don't panic just yet, because NASA scientists are on the case. They may have just discovered this space rock, but they're already tweeting about it like it's no big deal. "Oh, it's just a tiny asteroid, no need to worry," they say. Famous last words, am I right?

Now, let's get one thing straight - this asteroid may be small, but it still packs a punch. It's only 165 feet across, which is basically like a speck of dust in the grand scheme of things. But if it were to crash into a major city or densely populated area, it could still cause some serious damage. So maybe don't go planning that romantic picnic in the park just yet.

But fear not, fellow earthlings! NASA has got our backs. They've been working hard to develop methods to protect us from potentially hazardous asteroids like this one. Recently, they pulled off a pretty impressive feat - they managed to alter the trajectory of a small asteroid as an experiment by slamming a spacecraft into it. Yeah, you heard that right. Take that, space rock! By changing the trajectory, they could, maybe, avert the potential disaster caused by an earth-bound asteroid. Follow-up missions are already in the works to make this planetary defense technique even more effective. Just let NASA handle it. After all, what could possibly go wrong?

Comets

So, there are these things called comets, and they're like these icy space nuggets that come from way out yonder in the Kuiper Belt, which is like billions of miles away from us. And get this, some of these bad boys come from even farther out in the Oort Cloud, which is like six times farther than that! Yeah, and the reason comets have tails is that they get heated up by the sun and start to melt. And then, like a rocket ship, they shoot out gas and dust, which creates that funky tail we see. So, next time you're looking up at the sky and see a comet, remember that it's a fascinating space rock that's been around since the beginning of time, and we're lucky to witness it. Comets are fascinating celestial objects that have captured the imagination of people for centuries. They provide important clues about the formation of our solar system and are studied by astronomers to understand their behavior and evolution better.

Comet Hartley 2 has been visited several time by our space probes

Comets come in all shapes and sizes, from tiny ones that you could miss if you blinked to massive ones that are like miles wide. They're like that one friend who's always changing their mind about what they wanna do. And they like to travel around the sun in these wacky, elliptical orbits, which means they never quite follow the same path twice. The speed at which comets travel varies depending on their distance from the Sun. As they approach the Sun, their speed can reach up to tens of miles per second due to the enormous gravitational pull.

But don't worry, these space snowballs aren't really a threat to us. Most of the time, they steer clear of our planet. But, there's always that chance that one could come crashing down on us, and if it's big enough, we'd be in trouble. OMG!

Human-Made Satellites

Alrighty, let's talk about some space history. Back in 1957, the Soviets launched this funky metal ball called Sputnik 1 into space, which was the first-ever manufactured satellite. So, there you have it - rockets are the true rockstars of the space industry, and they're the reason we can explore the final frontier.

It hung out up there for a few months before doing a graceful swan dive back to Earth. But let's be real here, the true stars of early space exploration were the apes! The United States and the Soviet Union both sent these hairy creatures to space in the 1960s, and they were swinging their way to fame. Ham, the chimpanzee, even made it to an altitude of 157 miles before safely returning to Earth. Talk about a true space chimp-ion! And by the way, they also sent dogs up there too.

Fast forward to today, and we've got over 3,600 satellites orbiting the Earth. They're like cosmic butlers, providing all sorts of services like communication, navigation, mapping, and even spying on other countries. Yeah, you heard me right; they're like interstellar peeping toms.

So, maybe you're asking if all those fancy satellites circling our planet go around in the same direction. Yes! They do! And that direction is counterclockwise, in case you were wondering.

But why, you ask? It's simple, really. You know how you give a good push to a merry-go-round to make it spin faster? Well, it's the same thing with satellites! We launch those suckers into space, and thanks to Earth's rotation, they get a little speed boost that sends them on their merry way. And since we humans like to keep things simple, we send most satellites in a prograde orbit, which just means they follow the Earth's rotation.

Now, here's the fun part. There are a few rebels out there - I'm talking about those retrograde orbiters. They're like the bad boys of space, going against the flow and doing things their own way. But you know what they say, you gotta be a little crazy to change the world! These retrograde satellites are usually doing some top-secret science-y stuff or spying on other countries (shh, don't tell anyone I said that). So, while the prograde satellites are just chillin' and enjoying the ride, the retrograde ones are out there causing trouble and breaking all the rules.

The rockets that launch the satellites are no joke - they've come a long way since the days of Sputnik 1. Nowadays, we've got all kinds of rockets to choose from, ranging from sleek and sexy to big and beefy.

The secret to a rocket's success is all in the fuel. Most rockets use a mix of liquid and solid fuels that burn hot and fast, generating enough thrust to blast off the launch pad and escape Earth's gravitational pull. Liquid fuel rockets are like the fancy sports cars of the rocket world - they use liquid oxygen and liquid hydrogen or kerosene to power their engines. Solid fuel rockets, on the other hand, are more like muscle cars - they're packed with a mixture of fuel and oxidizer that burns hot and fast, propelling the rocket skyward.

One of the biggest players in the rocket game is SpaceX, and their Falcon 9 rocket is a real crowd-pleaser. It's got sleek lines and a reusable design that lets the first stage land back on Earth like it's no big deal. It's like the James Bond of rockets - smooth, sophisticated, and always ready for action.

Then you've got the Atlas V, (five) which is like the Hulk Hogan of rockets - big, beefy, and ready to launch anything you throw at it. It uses a combo of liquid oxygen and kerosene fuel, and can handle some seriously heavy payloads, like satellites and scientific probes.

But launching a satellite into orbit is no easy feat, my friend. It takes some serious brainpower and precision calculations to make sure everything goes smoothly. Once the rocket reaches the right altitude and speed, the satellite is released from the upper stage, and voila! It's on its way to orbit around the Earth.

The International Space Station

The space station is like a giant flying house in space, complete with living quarters, laboratories, and all sorts of high-tech gadgets. And yes, it's pretty darn big! So, there you have it, folks - the International Space Station dream team and their out-of-this-world experiments. These countries may speak different languages and have different cultures, but when it comes to exploring space, they're all on the same team. It's like a cosmic soccer match but with way cooler uniforms.

It is actually one of the largest structures ever built in space. It weighs over 441 tons and is roughly the size of a football field. That's a whole lot of space to play around in!

But here's the crazy part: scientists managed to get this massive structure into orbit around the Earth using rockets - they launched the individual pieces of the space station into space and then used robotic arms and spacewalking astronauts to assemble it piece by piece.

It was like putting together a giant space puzzle, but with the added challenge of microgravity and the harsh environment of space, like exposure to extreme heat and cold cycling, ultrahigh vacuum, atomic oxygen, high-energy radiation, and debris impact. The process took years. It's like a giant space Lego set, made up of multiple modules that can be added and removed as needed.

And once the space station was fully assembled, it was like a giant floating laboratory. Scientists from all over the world could conduct experiments in fields like biology, physics, and astronomy, all while orbiting high above the Earth.

Picture this: it's the 1980s, the Cold War is in full swing, and the United States and the Soviet Union are locked in a fierce space race. They've both got plans to build their own space stations, but then they have a sudden moment of genius: why not team up and build one together? It's like the ultimate buddy cop movie but in space!

And that's exactly what they did. Along with some help from Canada, Japan, and Europe, they created the International Space Station. It's like a giant science fair project that's out of this world! The first piece of the station was launched in 1998, and it's been growing ever since. It's got everything from living quarters to labs to systems that keep everything running smoothly. It's basically a floating city, but without the traffic jams and annoying neighbors.

But why did they build this giant space station in the first place? It's all about science, baby! The station is a platform for conducting experiments in fields like biology, physics, and meteorology. They're studying everything from the effects of zero gravity on the human body to developing new technologies for exploring other planets. It's like a big laboratory, but with an epic view of the cosmos.

But it's not just about science; it's also about international cooperation. The International Space Station is like a beacon of peace and harmony up there in the sky. It's like a little oasis of love and unity, a place where countries that may have had differences on Earth can come together and work towards a common goal. It's like a cosmic United Nations, but with cooler suits.

And who are the cool cats working up there? It's like the Avengers of space exploration, with NASA (United States), Roscosmos (Russia), JAXA (Japan), ESA (Europe), and CSA (Canada) all bringing their A-game. Each country brings its unique flavor to the space station, and its contributions have been crucial to its success.

For example, NASA has been sending up mission after mission to deliver equipment and materials to the station, while Roscosmos has been responsible for launching crew members and keeping the station's systems in tip-top shape. JAXA has been working on developing new technologies for future missions to the moon and Mars, while ESA has been studying the Earth from above. And let's give it up for CSA and their super cool robotic arm that moves equipment around the station. It's like having your very own Iron Man suit, but with less risk of getting blown up by a villain.

But what about the experiments they're conducting up there? NASA has been growing plants in space, which is not only super cool but could also be useful for future long-term missions. Roscosmos has been studying the effects of zero gravity on the human body, which could help us figure out how to keep astronauts healthy during extended trips. JAXA has been testing a prototype lunar rover on the station, which is like a tiny Mars rover that can hover in mid-air. And ESA has been studying the Earth from above, which is not only fascinating but also helps us understand how to better take care of our planet.

So, there you have it, folks - the International Space Station dream team and their out-of-this-world experiments. These countries may speak different languages and have different cultures, but when it comes to exploring space, they're all on the same team. It's like a cosmic soccer match but with way cooler uniforms.

The Hubble Telescope

The Hubble Space Telescope is the ultimate diva of telescopes – it's got a name that everyone knows, and it's been killing it since it was launched by NASA in 1990. It is the Beyonce of telescopes, and it's always slaying. This baby is high up in the sky, circling the Earth every 96-97 minutes at a blazing 17,000 miles per hour – it's like it's on a never-ending world tour!

The Hubble is always looking its best too – it doesn't get affected by the Earth's atmosphere like those other telescopes down below. This means that it can snap pics of distant objects without any distortion, and let me tell you, it's got the best angles in the business.

At 43 feet long and weighing in at a whopping 12.5 tons, the Hubble's got curves in all the right places. It's got an 8-foot primary mirror that's basically like a massive reflector for all the light from the universe to bounce off of. And it's got a whole bunch of instruments that analyze all that light and tell us what's out there.

The Hubble's had a few touch-ups over the years to make sure it's always on fleek. In fact, it got a total makeover in 2009, which basically made it the telescope equivalent of getting a whole new wardrobe.

This diva's not going anywhere anytime soon either – it's got a schedule booked out until the 2030s, and it's not slowing down. The Hubble's already made some major discoveries, like finding out about dark energy and spotting planets that aren't in our solar system. It's like the telescope equivalent of the queen of pop – inspiring a whole new generation of scientists and astronomers.

But don't think for a second that Hubble is the only star in the sky. The Chandra X-ray Observatory telescope, also operated by NASA, is all about those X-ray emissions from cosmic sources like black holes and supernova remnants. And the Spitzer Space Telescope, also operated by NASA, is all about that infrared radiation, dude. It's perfect for checking out dusty regions of space and exoplanets.

And the new kid on the block - the Sentinel-6 Michael Freilich. Launched in 2020, it's all about observing sea level changes and ocean currents around the globe. It's like the Instagram influencer of space telescopes - always looking out for that perfect angle.

The James Webb Space Telescope

I'll bet you want to know about the Rolls Royce of space telescopes – the James Webb Space Telescope. Well, let me tell you, it's a real doozy of a space telescope.

First off, let's talk about its size. It's huge! I mean, we're talking about a telescope that's the size of a tennis court. That's one big piece of equipment! It's like the Shaquille O'Neal of space telescopes - big, powerful, and ready to dunk on the competition. It's a big step up from the Hubble Space Telescope.

How do you suppose it got into space? They launched it on Christmas Day! Can you believe that? It's like NASA said, "Merry Christmas, universe! We've got a present for you!" I bet Santa was pretty jealous he didn't get to ride on that rocket.

The Jame Webb Telescope was launched on Christmas Day, 2022

So, where is it now? It's hanging out at a point in space called La Grange Point 2, which is about 1 million miles away from Earth outside the orbit of the moon. It's like the cool kid at the space party, just hanging out in the VIP section and checking out all the cosmic shenanigans.

And what is it capable of seeing? Well, let me tell you, this super-duper instrument can see pretty much everything. It's like having the world's best pair of glasses but for space. It can see back in time to when the first galaxies were forming, study exoplanets, observe star formation, supernovas, black holes, and basically any other space thing you can think of. It's like the Swiss Army Knife of space telescopes - it can do just about anything.

But why is it important for us to see distant galaxies? Well, let's put it this way - it's like reading a history book about the universe. By studying these galaxies, we can learn about the early days of the universe and how it evolved over time. It's like we're cosmic detectives trying to piece together the mystery of the universe's past. Plus, who knows, maybe we'll find some space treasure out there. I'm talking about alien life, people! It's like searching for the ultimate Easter egg in the universe.

The Webb telescope has been busy playing detective and looking for atmospheres on some rocky Earth-size exoplanets. And let me tell you, these planets are no easy catch. One of the planets in question is called LHS 475b or GJ 4102b, which sounds more like a code name for a top-secret government project than an actual planet. But hey, I'm not here to judge.

Let me explain briefly how scientists detect a planet orbiting a sun many light years away. They use a few different methods like watching a wobble of the distant sun and the planets caused by the gravitational affect each other on the computer screen or a dark image of the planet on its transit in front of a bright sun.

Anyways, NASA's primary planet-hunter, the Transiting Exoplanet Survey Satellite, had already caught wind of a potential planet, but it was up to Webb's scope to confirm its existence. And boy, did it ever. It only took two transits of the exoplanet in front of its star for Webb to detect it, living up to its fame of being the largest and most powerful space observatory to date. Take that, Hubble!

But the real kicker is what Webb found out about this planet's atmosphere. It's not dominated by hydrogen, like the gas giants in our own solar system - and a methane-rich atmosphere like the one on Saturn's moon Titan? Not likely, folks. Webb's researchers determined this based on how much starlight the planet blocked during its transits. It's like they're playing a game of "guess the atmosphere," but with way cooler equipment than we have on Earth.

This planet could potentially have an atmosphere rich in carbon dioxide, like Venus. And let me tell you, if this planet is anything like Venus, we should probably stay away. Venus is hot, like really hot. Like hot enough to melt lead. And Webb's observations indicate that this planet is a "few hundred degrees" warmer than Earth. Wow!

So, the James Webb Space Telescope is one big, cool, and powerful space telescope that's ready to explore the universe. It's like the Chuck Norris of space telescopes - it can kick cosmic butt and take cosmic names. And who knows, maybe one day we'll be able to high-five an alien thanks to this guy.

Space Junk

Picture this: a junkyard in space! Yep, that's right, it's called space junk. It's all the trash that we humans have sent up there over the years - old satellites, rocket parts, and even tiny bits of debris from collisions and explosions. And let me tell you; it's a real problem. All that junk is just floating around up there, wreaking havoc on our space missions. There's always a risk of a cosmic fender bender. It's like bumper cars but in space! So, space agencies and satellite operators have to carefully monitor their orbits to avoid any smash-ups. However, with more and more space debris floating around, the chances of collisions are higher than a giraffe's neck.

I mean, imagine trying to dodge a discarded toaster at 17,500 miles per hour! That's like trying to play dodgeball with a rocket launcher. Not cool. And it's not just dangerous for astronauts and spacecraft; it can also mess with our communication systems and scientific observations of space.

In fact, a collision between a Russian and U.S. satellite in 2009 created a massive cloud of debris that threatened other satellites in the area. It was like a game of dodgeball, but instead of rubber balls, it was space junk. So, it's time for some cosmic spring cleaning to reduce the amount of space debris and prevent future collisions. We need to get our space brooms out and start sweeping away all that space dust. Who knows what kind of treasures we might find hidden underneath all that debris? Maybe a long-lost sock or a sandwich that some astronaut accidentally dropped. The possibilities are endless!

But don't worry too much, scientists and engineers are working hard to come up with ways to deal with space junk. From giant nets to laser beams, they're trying it all. Who knows, maybe one day we'll be able to clean up that cosmic trash heap and turn it into a cosmic treasure trove.

Chapter Twenty-One

The Exciting Quest for Exoplanets

Picture this: it's October 1995, and Michel Mayor and Didier Queloz are just two astronomers peering up at the stars from Europe. Suddenly, they make a mind-boggling discovery - they spot the first planet outside our solar system! But this isn't just any old planet, folks. It's a "hot Jupiter" - a gas giant exoplanet that's 51 light years away from Earth. And they name it...wait for it... Dimidium. It sounds like a fancy Latin name for a rare plant species, but hey, it's the name they chose.

Of course, the Americans couldn't let the Europeans hog all the glory, so the folks over at San Francisco State University decided to get in on the exoplanet action too. Armed with their trusty Lick Observatory, they set out to find more exoplanets. And boy, did they succeed! In a planet-hunting frenzy that lasted from 1995 to 2005, they found a whopping 70 out of the first 100 exoplanets. Now that's what you call a winning streak.

So, how did they find all these planets? Well, it wasn't easy, let me tell you. They had to patiently stare through a giant telescope and watch as starlight wiggled and jiggled, thanks to the gravitational pull of an orbiting planet. It's like watching a cosmic ballet dance with stars and planets instead of ballerinas and tutus. And to detect these tiny changes in starlight, they used a technique that detects a Doppler shift in the host starlight. In other words, they were looking for signs that the star was wobbling ever so slightly due to the gravitational effects of a nearby planet.

But not just any planet would do - they were looking for giant, beefy exoplanets that orbited close to their host star. Think of it like a big bully pushing a smaller kid around on the playground. Except in this case, the bully is a planet, and the kid is a star. And the exoplanet hunters were the ones who caught the bully in the act.

So next time you look up at the stars, remember - there might be a whole bunch of hot Jupiters and cosmic bullies out there, just waiting to be discovered.

Some of the Exoplanets

In simple terms, an exoplanet is a planet that orbits a star other than our Sun. In other words, it's an interstellar traveler who has found a new home away from their original star. These planets come in all shapes and sizes, from rocky Earth-like planets to gas giants the size of Jupiter.

Astronomers have been on the lookout for these bad boys for the past few decades, and they've discovered thousands using a variety of detection methods.

Let me give you some examples of the coolest exoplanets out there so far discovered:

First off, there's **Kepler-186f** - the first confirmed Earth-sized exoplanet discovered within its star's habitable zone. It orbits a red dwarf star about 500 light-years away from us, so you know it's a little out there. That means it's in a sweet spot where it could have liquid water on its surface.

Then there's **TRAPPIST-1**e, one of seven planets orbiting a red dwarf star called TRAPPIST-1. It's similar in size to Earth and located in the star's habitable zone, making it a good candidate for supporting life. Who knows, maybe some tiny alien creatures are partying it up on this planet right now.

Next up, we've got **HD 209458b**, a gas giant exoplanet located about 150 light-years away from us. It was the first exoplanet to have its atmosphere directly detected, and it's got some seriously interesting stuff going on - think water vapor, carbon dioxide, and other funky chemicals.

Then there's **WASP-12b**, a gas giant exoplanet that's orbiting way too close to its star-like, completing one orbit every 1.1 days. It's so close that its atmosphere is being stripped away by the star's radiation, creating a massive cloud of gas around the planet. Talk about being in a hot spot!

Last but not least, we have **51 Pegasi b** - the first exoplanet discovered orbiting a sun-like star. This was a major breakthrough in exoplanet research, and it's a gas giant exoplanet that orbits super close to its star, completing one orbit every 4.2 days.

Discovering exoplanets is like a never-ending cosmic treasure hunt, and astronomers are always on the lookout for new loot. It's like trying to find Waldo, but he's a planet, and he's hiding somewhere in the vast expanse of space.

And let's be real, with the number of stars in our galaxy alone; there are probably more exoplanets out there than grains of sand on a beach. We need to keep looking, sifting through the cosmic sand to find those elusive little rascals.

Who knows what kind of exoplanets we'll discover next? Maybe we'll find a planet made entirely of chocolate or one with giant flying unicorns. Okay, perhaps not those exact ones, but you get the point. The possibilities are endless, and that's what makes exoplanet hunting so exciting.

Some of these exoplanets are even similar in size and composition to our good old Earth, and some even orbit their stars within the "habitable zone," where it's not too hot and not too cold for liquid water to exist. That means they could potentially have life on them! We could be talking about little green (or purple or yellow or polka-dotted) aliens!

And the best part? We've only just scratched the surface. There are still so many exoplanets out there waiting to be discovered, and who knows what secrets they might hold.

Chapter Twenty-Two

Is Anybody Out There? The Search for Extra-Terrestrial Life

Hey, let's talk about the ultimate search that has everyone buzzing - the hunt for extraterrestrial life! It's like the ultimate cosmic game of hide-and-seek, and scientists have been playing it for centuries. They're like the nerdy version of Ghostbusters, but instead of ghosts, they're looking for aliens. And let's be real, who wouldn't want to find some crazy, space-dwelling creatures? It's like discovering a new species of animal, but on steroids! We could be talking about a universe teeming with life, or we might just be stuck with ourselves, holding down the fort as the only intelligent life form in the entire universe. That's a pretty lonely thought, right? But fear not; the SETI Institute or the Search for Extraterrestrial Intelligence is on the case!

SETI sounds like something straight out of a sci-fi movie, but these guys are the real deal. These brilliant minds have been searching for aliens since 1984, using all sorts of techniques from listening for signals from other planets to looking for chemical signs of life on distant worlds. They even have fancy telescopes to scan the sky for any signs of ET phone home. Who knows, maybe one day they'll receive a signal from an alien civilization that says, "Greetings, Earthlings!" and we'll all be invited to a wild space party. Count me in!

So, if you're into space and the possibility of finding intelligent life beyond our planet, the SETI Institute is the place to be. They're like the ultimate cosmic detectives, searching for clues to answer the age-old question: are we alone in the universe? As the renowned astronomer Carl Sagan once said – "The Universe is a pretty big place. If it's just us, seems like an awful waste of space."

The possibilities are endless, and the search for alien life has given us an incredible understanding of our own place in the cosmos. We're learning about our origins, how life began, and what could be out there. And who knows, maybe one day we'll find a planet just like Earth, with oceans, clouds, and a breathable atmosphere, and maybe even some chill alien neighbors! It's like finding a needle in a haystack, but with a magnifying glass and some high-tech gear, we just might find it. So, what are we looking for exactly? Well, first things first, we need a planet with liquid water, which is essential for supporting life as we know it. And it has to be in the Goldilocks zone of its star system - not too hot, not too cold, just right, baby! We're talking about the perfect balance of gases in its atmosphere, too. That way, we might find a range of life forms, just like we have here on our planet. No pressure, right? The search for extraterrestrial life is a tough job, but someone's gotta do it. And if we can find that one little world that's just right, it will be a game-changer! It's a thrilling prospect and one that fuels our curiosity, determination, and imagination. We may not know exactly what we're looking for, but we know it's out there, and we won't stop until we find it. So, let's keep exploring, fellow space cadets, and who knows what we'll find!

Unidentified Flying Objects

Oh boy, we're talking about UFOs! You know what UFO stands for, right? It's an Unidentified Flying Object, which means any object or light in the sky that we can't immediately identify. UFO sightings have been reported for many years, and they continue to fascinate and intrigue people all around the globe.

And let me tell you, these things can take all sorts of shapes and sizes! We're talking about bright lights, disc-shaped objects, and even triangular or cylindrical shapes! It's like a real-life game of "what's that thing in the sky?" Now, a lot of people associate UFOs with aliens, and while that's definitely a possibility, we can't say for sure what they are or where they come from. That's why some folks have started using the term UAP - Unidentified Aerial Phenomenon - to describe these mysterious sightings. It's like UFO's more sophisticated cousin, you know? And apparently, some government agencies and military branches are all about it. They want to investigate these sightings and get to the bottom of this whole thing. The idea is that by using a more neutral term like UAP, we can reduce the stigma that's often associated with UFOs. You know, all that pseudoscience and conspiracy theory stuff. We're talking about the real scientific investigation here, people! And who knows? Maybe we'll find some evidence of extraterrestrial life along the way. Wouldn't that be wild? But until we have concrete evidence, the debate rages on. Are these sightings just natural phenomena or misidentifications? Or are we really being visited by beings from another planet? The truth is out there, folks - and it's up to us to figure it out! And maybe, one day, we will. I am an eternal optimist!

Rocket Man

Becoming an astronaut is not as easy as you might think. First things first, you're going to need a degree in STEM. And no, that doesn't stand for "Sipping Tea and Eating Muffins" - it stands for science, technology, engineering, and math. So, if you're not a fan of math or science, you might want to consider another career, like underwater basket weaving.

But it's not that easy - you'll also need flight experience. That means you'll have to spend some time in the cockpit, flying planes and getting your wings. And don't even think about faking your way through it. The last thing we need is an astronaut who thinks "autopilot" means they can take a nap.

Of course, you'll need to be in top physical shape as well. And no, eating kale for a week straight doesn't count. You'll need to hit the gym and get those muscles pumping. Because let's face it, if you can't bench press a rocket, you're not going to make it to space.

But it's not just about your body - you'll need to be smart, too. And not just "I know all the words to 'Rocket Man'" smart. You'll need to have experience in a technical field, like engineering or computer science. And if you've got experience in a space-related field, like alien communications or space archaeology, all the better.

Oh, and don't forget about language skills. You'll need to be fluent in English, and maybe even learn a few more languages, like Klingon or Ewok. Because you never know when you might encounter some extraterrestrial life forms.

And last but not least, you'll need to be mentally tough. That means being able to handle long periods of isolation and working with a team of people who are just as crazy as you are. Because let's face it, you've got to be a little bit nuts to want to blast off into space.

So, there you have it - everything you need to become an astronaut. It's not easy, but hey, if you can handle rocket science, you can handle anything, right? Just remember to pack your sense of humor, because you're going to need it when you're hurtling through space at thousands of miles per hour. And who knows, maybe you'll be the first astronaut to discover a new planet or make contact with an alien species. Just don't forget to send us a postcard from the stars.

Captain Kirk

William Shatner is a Canadian dude who's got a whole lot of talents up his sleeve. He's an actor, author, producer, director, and singer, and probably knows how to cook mean poutine too (or maybe not, who knows?). But what he's most famous for is playing the legendary Captain James T. Kirk in command of the USS Enterprise in the TV series, Star Trek - you know, that show with the pointy-eared aliens and space travel shenanigans.

As captain of the USS Enterprise, Captain Kirk was dedicated to exploring "strange new worlds," seeking out "new life and new civilizations" and boldly going "where no one has gone before."

Well, turns out he's not just a sci-fi hero on-screen - he's also a real-life space explorer! In 2021, he got to ride in the Blue Origins space shuttle and venture into the final frontier.

Yeah, he got to go to space on a real-life spaceship, not just a fancy Hollywood set. And apparently, while he was up there, he did some real astronaut stuff, like looking out the window of the spacecraft.

Shatner's first reaction to being in space was...not exactly what you'd expect. Apparently, the guy who boldly went where no man has gone before was feeling pretty darn dark and gloomy.

Instead of staring lovingly at our beautiful planet Earth as most people would, he looked in the opposite direction. That's right; he was staring off into space like a moody teenager who just got grounded.

In his book, he wrote, "I love the mystery of the Universe. I love all the questions that have come to us over thousands of years of exploration and hypotheses. Stars exploding years ago, their light traveling to us years later; black holes absorbing energy; satellites showing us entire galaxies in areas thought to be devoid of matter entirely... all of that has thrilled me for years..."

However, as he looked out the window of the spacecraft and looked in the direction opposite Earth, he wrote, "there was no mystery, no majestic awe to behold. All I saw was death. I saw a cold, dark, black emptiness. It was unlike any blackness you could see or feel on Earth. It was deep, enveloping, all-encompassing."

As he turned back toward what he described as "the light of home," he saw the opposite. "I could see the curvature of Earth, the beige of the desert, the white of the clouds, and the blue of the sky. It was life. Nurturing, sustaining, life. Mother Earth. Gaia. And I was leaving her."

Then he had a stunning revelation: "Everything I had thought was wrong. Everything I had expected to see was wrong."

"It was among the strongest feelings of grief I have ever encountered. The contrast between the vicious coldness of space and the warm nurturing of Earth below filled me with overwhelming sadness. Every day, we are confronted with the knowledge of further destruction of Earth at our hands: the extinction of animal species, of flora and fauna . . . things that took five billion years to evolve, and suddenly we will never see them again because of the interference of mankind. It filled me with dread. My trip to space was supposed to be a celebration; instead, it felt like a funeral."

Conclusion

The universe, folks, is a big, big place. Like, seriously massive. We're talking billions and billions of galaxies, each containing billions and billions of stars and who knows how many planets, moons, asteroids, and other space doodads. It's enough to make your head spin. So, my fellow space folks, this is where this part of the voyage ends!

But that's not all. No, sir, this universe has some serious mystery and intrigue. We're talking about dark matter and energy, which sounds like something out of a sci-fi movie but is actually real (or so the scientists tell us). These invisible forces make up most of the universe, but we have no idea what they are or how they work. It's like trying to solve a Rubik's cube blindfolded while riding a unicycle. Good luck with that.

Then there's our solar system, which is pretty cool if you think about it. We've got the Sun, which is basically a giant flaming ball of gas that we rely on for warmth and light (thanks, Sun!). And orbiting around it, we've got eight planets (or nine, depending on whom you ask), each with its own unique personality. There's Mercury, which is so close to the Sun that it's practically on fire. Venus has a thick, toxic atmosphere that would make Darth Vader feel right at home.

Earth is our home sweet home, and the only planet we know of that has beer (probably). Mars has a reputation for being a bit of a loner, but we're still rooting for it. Jupiter, which is so big could swallow up all the other planets and still have room for dessert. Saturn has some serious bling in the form of its rings. Uranus, which has a name that still makes us giggle. And Neptune, which is so far out there, we're surprised it hasn't gotten lost.

Of course, we can't forget about all the other space stuff. There are asteroids, which are basically giant space rocks that could wipe out humanity if they ever decided to pay us a visit (please don't). There are comets, which are like dirty snowballs that leave a trail of ice and dust behind them as they travel through space. And then there are all the other space objects that we're still trying to figure out, like dwarf planets and trans-Neptunian objects. We're not sure what they do, but they sound important.

And let's give tribute to the two Voyager spacecraft, which is basically the coolest interstellar road trip ever. They've been traveling through space for over 40 years, sending back photos and data from the far reaches of our solar system. It's like the ultimate Instagram influencer, except it doesn't care about likes or followers.

But perhaps the most exciting thing about the universe is the search for exoplanets. These are planets orbiting stars other than our Sun, and we've found thousands of them. Some might even be capable of supporting life, which is like finding a needle in a haystack (or a beer in a desert). It's got scientists and astronomers all giddy with excitement, like kids on Christmas morning.

All in all, the universe is a pretty amazing place. It's full of mysteries, wonder, and a whole lot of weirdness. We may never know all the answers, but that's okay. The journey of discovery is half the fun.

But at this point, you may be asking this question - Why the heck is there a universe? It's the biggest question of all time, and even the smarty-pants scientists, philosophers, and theologians can't agree on a definitive answer. Some say it all started with a big bang about 13.8 billion years ago. Others reckon some higher power whipped it up for a laugh. And then there are those who say the universe is just here for us to create our own meaning in it.

But honestly, who knows? Maybe the universe is just a cosmic prank that we'll never figure out.

You remember the Voyager 1 spacecraft? It had snapped a picture of the blackness of space and in that picture was an isolated little dot, appearing against the black backdrop of the vastness of space. It became famously known as the *Pale Blue Dot*.

The renowned popular astronomer, Carl Sagan wrote this... "Look again at that dot. That's here. That's home. That's us. On it, everyone you love, everyone you know, everyone you ever heard of, every human being who ever was, lived out their lives. Our posturings, our imagined self-importance, the delusion that we have some privileged position in the Universe, are challenged by this point of pale light. There is perhaps no better demonstration of the folly of human conceits than this distant image of our tiny world. To me, it underscores our responsibility to deal more kindly with one another and to preserve and cherish the pale blue dot, the only home we've ever known."

So, my fellow space folks, this is where this part of our voyage ends!

CONCLUSION

Our little Pale Blue Dot called Earth

Take a Minute to Share the Knowledge!

Maybe we'll never figure out why the universe exists... and maybe it's better that we don't. But what we can do is learn as much about it as we can, and spread that knowledge far and wide.

Simply by sharing your honest opinion of this book on Amazon, you'll show new readers where they can find this fascinating information, all packed into one easy-to-digest morsel.

LEAVE A REVIEW!

Thank you so much for your support. I'm honored to share this universe with you!

>>>Please be so kind as to leave a review using this link...

amzn.to/45j5ICY

About the Author: Oscar Joe Washington

Oscar Joe Washington is a debut author whose love for science and astronomy inspired his first book. Born and raised in a suburb of Detroit, Michigan, Oscar is a Millennial child who grew up in a multicultural family. His parents, originally from Kenya, migrated to the USA in the mid-1970s and settled in Michigan. In his mid-teens, the family moved to Los Angeles, California, where Oscar pursued a degree in science at a college.

From a young age, Oscar had a fascination with the night sky, which was sparked by his father's passion for amateur astronomy. His father would often take him stargazing, and this led him to read books on the subject and attend seminars and workshops. This early exposure to astronomy and astrophysics played a significant role in shaping Oscar's interests and later inspired his writing.

Oscar's debut book is a science fiction story that explores the mysteries of the universe. He weaves scientific concepts and theories into the plot, drawing on his knowledge and passion for science. The book follows the journey of a young protagonist who embarks on a quest to unravel the secrets of the cosmos. His book offers a fresh perspective on science fiction and invites readers to contemplate the mysteries of the universe.

When he is not writing, Oscar enjoys stargazing, hiking, and exploring the outdoors. He is passionate about science education and is committed to inspiring the next generation of science enthusiasts. He is a firm believer in the power of science and technology to make the world a better place.

Acknowledgements

HTTPS://WWW.UNIVERSETODAY.COM/24587/BLUE-GIANT-STAR

HTTPS://IMAGINE.GSFC.NASA.GOV/SCIENCE/OBJECTS/DWARFS2.HTML

HTTPS://WWW.SPACE.COM/22180-NEUTRON-STARS.HTML

HTTPS://WWW.SPACE.COM/32661-PULSARS.HTML

HTTPS://EARTHSKY.ORG/SPACE/DEFINITION-WHAT-ARE-BROWN-DWARFS

HTTPS://SOLARSYSTEM.NASA.GOV/SOLAR-SYSTEM/SUN/OVERVIEW

HTTPS://WWW.SPACE.COM/58-THE-SUN-FORMATION-FACTS-AND-CHARACTERISTICS.HTML

HTTPS://WWW.SPACE.COM/22437-MAIN-SEQUENCE-STAR.HTML

HTTPS://SCITECHDAILY.COM/18-MILLION-MILES-PER-HOUR-STAR-WITH-THE-SHORTEST-ORBITAL-PERIOD-AROUND-BLACK-HOLE-DISCOVERED.HTML

HTTPS://WWW.SPACE.COM/40920-RELATIVITY-POWER-OF-EQUIVALENCE.HTML

HTTPS://WWW.SPACE.COM/LOOP-QUANTUM-GRAVITY-SPACE-TIME-QUANTIZED

HTTPS://WWW.SPACE.COM/35676-EINSTEIN-SPOOKY-ACTION-STARLIGHT-QUANTUM-ENTANGLEMENT.HTML

https://www.nbcnews.com/mach/science/what-relativity-einstein-s-mind-bending-theory-explained-ncna865496

https://www.bbcearth.com/news/what-would-happen-if-you-fell-into-a-black-hole

https://www.upworthy.com/william-shatner-shares-reflections-on-going-to-space

https://www.nasa.gov/feature/jpl/pale-blue-dot-revisited

https://positivethinkingmind.com/quotes-about-the-universe

HIGHEST HILL PUBLISHING

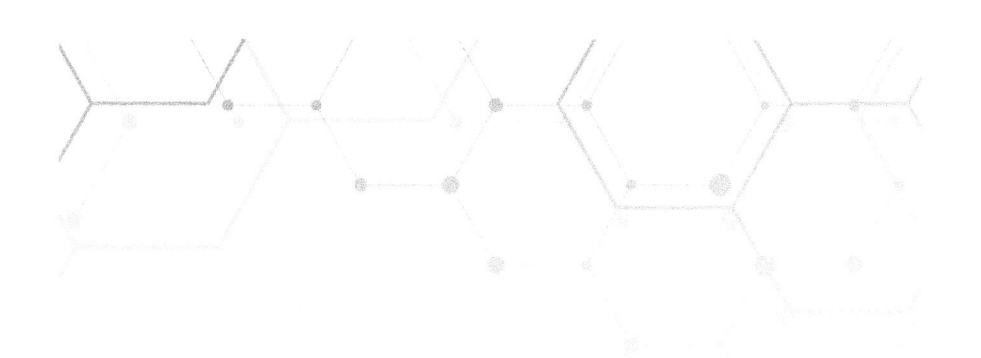

Milton Keynes UK
Ingram Content Group UK Ltd.
UKHW021342251024
2386UKWH00060B/943

9 798988 971603